Economic Growth vs. the Environment

Edited by:

Warren A. Johnson
San Diego State College

John Hardesty
San Diego State College

Wadsworth Publishing Company, Inc., Belmont, California

ISBN-0-534-00103-3
L. C. Cat. Card No. 77–171898
Printed in the United States of America
1 2 3 4 5 6 7 8 9 10—75 74 73 72 71

Preface

In a finite world no positive growth rate can be maintained forever; it is a physical impossibility. This principle, as applied to population growth, is rapidly gaining acceptance. It is the purpose of this volume to argue that an analogous conclusion regarding economic growth is also inevitable, and that current environmental conditions in the United States are serious enough to warrant consideration of ways to end our present dependence on economic growth.

This conflict between economic growth and quality of the environment is certainly not new, but it is only in recent years, with the conflict approaching crisis proportions, that written analyses have begun to appear. We hope that this volume will provide an initial steppingstone for discussing the problem of reconciling the economic needs of society with the quality of our environment. As such, it should be of interest to those seeking an interdisciplinary perspective reconciling the rapidly growing requisites of ecology with the social sciences—economics, geography, and particularly political science.

We are very grateful to those who have helped us draw these writings together. We especially want to thank our San Diego State colleagues Norris C. Clement of the economics department, Phillip P. Pryde of the geography department, and Gary Suttle, graduate assistant, for their contributions. We would also like to express our appreciation to Dr. Charles F. Cooper of the National Science Foundation, John Niedercorn, chairman of the economics department at the University of Southern California, and Daniel Luten of the geography department at the University of California at Berkeley. The editors, of course, assume complete responsibility for the final product and any errors or omissions that it may include.

Contents

Introduction

This book is an attempt to integrate into economic thinking what is presently known about ecology. Today, economics and ecology rarely interact. Economics is a mainstream concern, while ecology is peripheral and secondary. Economic factors such as unemployment, inflation, personal income, and individual opportunities are basic to the organization of our society, and hence to its health and vitality. These economic factors have tremendous political implications, and measures that might cause unemployment or place severe restraints upon an important industry would have little chance of approval. Only a small number of people would accept the primacy of ecological criteria over these economic criteria, and our society tends to see such people as taking ecology too far. Normally it is assumed that environmental requirements can be fulfilled without interfering with basic political and economic institutions. Even many conservationists feel that with proper planning it is possible to have increasing affluence and a high quality environment. The environmental movement has gained a great deal of sympathy for its point of view but has produced few major changes, possibly because it has not understood the full implications of its proposals on our economic society.

These contradictions in our attitudes toward man and nature are nothing new, however. They have been part of American culture almost from its start. Our sympathies were overwhelmingly with Thomas Jefferson's agrarian ideal; we had no taste for the gray industrial cities of Europe. But the agrarian ideal was only a myth, powerless to resist historic trends, for as a nation we chose rapid economic development and its inevitable urbanization. We honor Henry Thoreau as a man of letters, but we have paid scant heed to his admonition to reject our "desperate haste to succeed and in such desperate enterprises." Will the same be true for our ecological concerns of today? Will they be hopeless ideals in the face of the "practical" concerns of the day, the powerful economic and technological trends that are turning the earth's magnificent natural life-support systems into dangerous hazards, consuming irreplaceable resources and forcing an unwanted degree of urbanization? If it is different in the future it will be because the nature of the issue has changed. For Jefferson and Thoreau, their relationships with man and nature were a matter of choice. For us it is a matter of survival.

Ecologists, unfortunately, cannot tell us with certainty just what we need to know to enable us to cope with the ecological problems that are appearing so rapidly. Our understanding of these highly interrelated problems is already lacking, and the future can be expected to add new complexities. What type of technology will be developed? How resilient will natural systems be to continued increases in the scope and complexity of man's activities? How adaptable will man himself be to rapid social and environmental changes? All that ecologists can do is to make educated projections based on their study of natural ecosystems and certain of the more obvious environmental consequences of man's activities to date. On this limited basis, ecologists have pointed out a number of major hazards. Are they doomsayers or prophets? No one can say for sure without claiming visionary powers, but in one respect the message is clear: be careful with things we do not understand. Our future and the future of our children is at stake, and such high stakes suggest a cautious, deliberate approach.

Although ecologists cannot claim omniscience, they have developed a number of tentative principles about the functioning of living organisms and their nonliving environments—about ecosystems, to use the ecological term. When these principles are applied to our economic behavior it is hard to avoid the conclusion that present economic trends are diametrically opposed to the requirements of ecosystem stability. Instability, in ecology, generally means rapid population fluctuations, the die-offs and explosions of plant and animal populations that frequently follow man's intervention in the environment. Man, although highly adaptable and possessing powerful technologies, is still a biological organism and subject to the same sources of instability.

Economic activity seems to contribute to ecological instability in four major ways.

1. Ecosystem stability is enhanced by a diversity of species in the community, by a complex web of life. Factors of economic efficiency are taking us toward ecosystem simplification, toward monocultures—environments with only one species—whether it is specialized agriculture, animal husbandry, or forestry. In such ecosystems insects and diseases can spread rapidly and cause great destruction, as commonly occurs in poultry houses, corn and cotton fields, and commercial forests.

2. Cities can also be seen as monocultures of human populations, and therefore subject to instability. Economic forces make it very difficult for new economic enterprises to locate in nonurban places because of the limited transportation facilities of rural areas, the absence of large markets nearby for the product or service produced, and the limited supply of labor and raw materials. In other words, urban areas have higher economic

productivity, achieved through specialization of skills and technology, economic integration, and population concentration. The increased productivity gained through these processes, however, could prove to be disastrous if transportation was disrupted or supplies of water or energy were cut off. This has been expressed well by Robinson Jeffers:

> We have geared the machines and locked all together into interdependence; we have built the great cities, now
> There is no escape. We have gathered vast populations incapable of free survival, insulated
> From the strong earth, each person in himself helpless, on all dependent.

Any species that becomes too dependent on a specific set of conditions or requirements is violating the principle that stability is dependent on a complexity of interrelationships that offers alternative means of survival when other means disappear.

3. Ecologists caution against too great a transformation of the environment in which a species evolved. Man has been a remarkably adaptable species, but perhaps some of our physiological and psychological stresses stem from trying to stretch man's adaptability too far too quickly. Change today originates primarily in the application of new technology, which our economic system rewards so handsomely.

4. Our economic system maximizes output—Gross National Product—while ecology calls for balanced throughput, for limiting the consumption of resources to conserve the materials on which the functioning of the ecosystem depends and to avoid contaminating the ecosystem with an excess of waste. Recycling of materials is the key ecological process that accomplishes both of these objectives at once, turning waste into nutrients. Natural ecosystems accomplish this beautifully, and so do many forms of traditional agriculture. In urban industrial societies recycling is usually "uneconomical," or at best requires the use of additional energy and materials to accomplish it, so that the net ecological gain is significantly reduced.

Basically, it can be said that healthy natural communities are generally ones that are in *equilibrium* with their environment, while our economic system is dependent on *disequilibrium,* on rapid growth. As a result we are tormented with the necessity of having to adjust to all the change that inevitably stems from rapid growth. It is hard to escape the conclusion that the continuation of present patterns of economic growth will take us toward greater ecological instability and, sooner or later, into a period of chaos followed by the establishment of a balance at some lower level of population and productivity. The alternative that we must seek is to establish a *sustainable balance* with the environment. Progress and growth, the ideals that have been so much a part of the American dream, must be replaced by the ideals of balance and stability.

This premise will certainly not find easy acceptance. Economic growth, as a national goal, is probably accepted by more Americans than any other national objective. And even though many Americans feel threatened by the deterioration of the environment, both physically and socially, this fear is probably minor compared to the uncertainties involved in transforming our economic system into one that is ecologically rational and, therefore, profoundly different. Instead, we will hope that the enlightened self-interest of voters, legislators, and businessmen, coupled with technological breakthroughs, will be able to clean things up, to make the machine run clean. As President Nixon said in his 1970 State of the Union message:

> The answer is not to abandon growth, but to redirect it. For example, we should turn toward ending congestion and eliminating smog the same reservoir of inventive genius that created them in the first place. Continued vigorous economic growth provides us with the means to enrich life itself and to enhance our planet as a place hospitable to man.

Whether growth can be redirected is a question discussed at various points in this volume. Certainly, there is much that can be done to enable us to sustain an expanding economy for some time, perhaps quite a long time if we can assume that the participants in the process employ ecological common sense—a dubious assumption given the nature of current political and economic processes. Nor does redirecting growth resolve the other problems associated with the maintenance of an expanding economy, such as military spending, economic exploitation of less developed countries, and the poverty and racial injustice that accompanies the market in labor. Hence the purpose of this volume, to argue that we should begin to consider ways to control growth now while alternatives are still available and the environment is still habitable, and that we should not wait until we have driven ourselves into a dangerous corner in which the chances of avoiding ultimate catastrophe are greatly reduced.

Such are the concerns of this book—the economic trends of our society, their ecological implications, and the prospects for resolving not the environmental consequences so much as their origin in our economic dynamism and the type of technology it calls into use. Part 1 consists of several expressions of the ecological point of view, including the application of this view to our economic behavior. Part 2 is the core of the book, with arguments against economic growth from a number of points of view. The long-range ecological arguments are translated into the more immediate social and political questions that must be dealt with if present trends are to be modified. Part 3 presents a variety of arguments in support of continued economic growth. Several are subtle and sophisticated, the best efforts to resolve the problems of providing for growth that the editors could find. It is hoped that the reader will be able to evaluate these arguments with material presented in Parts 1 and 2. Part 4 is a parting shot at our reliance on economic factors to organize society. There are two strong challenges from the past that still have meaning for us today, and two contemporary declarations for the future.

1 The Ecologist's Perspective

Ecologists and economists, unfortunately, rarely talk with each other. Ecologists, it often seems, are reluctant to face the practical, self-assured economists and prefer to claim the fundamental rightness of their broad, long-run point of view and then stand back and wait for society to fall in line behind it. The economists, for their part, often dislike having to face the environmental implications of their craft, and much prefer the more pleasing vistas of abundant consumer goods which they have helped to provide. The lack of communication between the two professions is creating a dangerous situation.

Ecology is defined as the interrelations between living organisms and their environment. One of the best statements of the concept is also one of the earliest, that of John Muir in 1869: "When we try to pick out anything by itself, we find it hitched to everything else in the universe." It is the systems approach applied to the natural sciences, looking at entire communities and their physical environment rather than single species of plants or animals. If ecology is anything, it is holistic; it thus finds many different expressions. The first one presented here is that of the scientist, in "All about Ecology," by William Murdoch and Joseph Connell, two members of the biology department of the University of California, Santa Barbara. They discuss what ecology is today, the role of the ecologist, and why ecology truly is the "subversive science" in our society.

A very different expression of the ecological point of view is that of Ian McHarg. Trained as a landscape architect and now chairman of the Department of Landscape Architecture at the University of Pennsylvania, McHarg has explored ways in which ecology can be incorporated into principles of environmental design. His perspective is thus more that of the artist—bold, expansive, almost that of a new renaissance man who has emerged out of the dark ages of ecological ignorance. Using information from many scientific disciplines in his essay "Values, Process, and Form," McHarg describes the formation of the earth and its multitudinous forms of life which embody the evolutionary process toward greater diversity, stability, and interdependence, the process that has led to what McHarg refers to as the fitness of the environment. When man creates something that has this fitness, it is art. But this is rare, and McHarg rails at the "merchants' creed" of modern man that is radically transforming the results of evolution into something that is "unfit" and repellent to the human spirit.

Aldo Leopold was an ecologist before the word was hardly known; he was a natural scientist, but one with a predisposition to think of systems that include man. In the selection reprinted from "The Conservation Ethic," written in 1933, Leopold foresaw that conservation, if it was ever to be more than a cult, or political boosterism, would have to be part of a larger ethical system that transcends economics and science. Words such as "husbandry," "community," "love," and "respect for the land" appear throughout Leopold's writing. His classic *Sand County Almanac,* however, ends with the lament that in our treatment of the land and its myriad forms of life "We are remodeling the Alhambra with a steam shovel, and we are proud of our yardage." Leopold, in addition to making contributions to the philosophy of conservation that are comparable to Thoreau's and Muir's, made major practical contributions. He originated the wilderness program of the U.S. Forest Service and then in 1928 founded the profession of wildlife management at the University of Wisconsin.

The major sources of energy that power our American way of life, the fossil fuels (oil, gas, and coal), play a special role in the contemporary concern for the environment. Not only are they the major source of air pollution but they are also truly nonrenewable resources. Metals can be recycled; fossil fuels, once they are burned, have released as heat the chemical energy that was bound up in their molecular structure, and that cannot be recaptured. Our supplies of fossil fuels are limited—oil and gas especially. Nuclear energy, contrary to the popular conception, is far from being a reality as a long-range answer. Our standard of living is very closely tied to a very high level of energy consumption; without abundant energy, our life would be very different. Yet, as Garret de Bell points out, we are consuming our fossil fuels in exceedingly wasteful ways and at ever increasing rates, and environmental consequences are many. De Bell was a graduate student in biology at the University of California, Berkeley, until he dropped out of school to edit *The Environmental Handbook* and *The Voter's Guide to Environmental Politics.*

The last two selections involve the writing of Paul Ehrlich, the ecologist who has perhaps done more to bring ecology to a large audience than anyone else. The first is a review of Ehrlich's new book, co-authored with his wife, Anne, *Population, Resources, Environment.* The reviewer is Robert Heilbroner, a noted economic historian. He concludes the review with a fascinating discussion of what the terminus of the capitalistic system will be, whether it is to be a stationary state, according to John Stuart Mill, or a continuous, uncontrolled expansion as predicted by Karl Marx.

It is interesting that in *The Future as History,* written in 1959 Heilbroner does not mention the environment among the problems that our economy must come to terms with in the future. In this review, only eleven years later, he states that "I have slowly become convinced . . . that the ecological issue is not only of primary and lasting importance, but that it may indeed constitute the most dangerous and difficult challenge that humanity has ever faced."

The last selection is from Ehrlich's book itself, first a section on the extreme vulnerability to epidemics of our mobile and highly interdependent urban society, and a second in which he comments on economics and politics as "two sides of the same coin." Although Ehrlich has been criticized, as have most popularizers, his academic credentials are impressive. Starting as an assistant professor at Stanford University in 1959, he rose to a full professor and director of graduate study in the biology department by 1966, published the best-selling *Population Bomb,* and has published some 70 research papers. His energy is reflected in his style of writing; the arguments are fired off rapidly and convincingly.

William Murdoch and Joseph Connell: All about Ecology

. . . The average citizen is at least getting to know the word ecology, even though his basic understanding of it may not be significantly increased. Not more than five years ago, we had to explain at length what an ecologist was. Recently when we have described ourselves as ecologists, we have been met with respectful nods of recognition.

A change has also occurred among ecologists themselves. Until recently the meetings of ecologists we attended were concerned with the esoterica of a "pure science," but now ecologists are haranguing each other on the necessity for ecologists to become involved in the "real world." We can expect that peripatetic "ecological experts" will soon join the ranks of governmental consultants jetting back and forth to the Capitol—thereby adding their quota to the pollution of the atmosphere. However, that will be a small price to pay if they succeed in clearing the air of the political verbiage that still passes for an environmental policy in Washington.

Concern about environment, of course, is not limited to the United

---

Reprinted, by permission, from the January 1970 issue of *The Center Magazine,* Vol. III, No. 1, a publication of the Center for the Study of Democratic Institutions in Santa Barbara, California.

States. The ecological crisis, by its nature, is basically an international problem, so it seems likely that the ecologist as "expert" is here to stay. To some extent the present commotion about ecology arises from people climbing on the newest bandwagon. When the limits of ecological expertise become apparent, we must expect to lose a few passengers. But, if only because there is no alternative, the ecologist and the policymakers appear to be stuck with each other for some time to come.

While a growing awareness of the relevance of ecology must be welcomed, there are already misconceptions about it. Further, the traditional role of the expert in Washington predisposes the nation to a misuse of its ecologists. Take an example. A common lament of the socially conscious citizen is that though we have enough science and technology to put a man on the moon we cannot maintain a decent environment in the United States. The implicit premise here seems clear: the solution to our ecological crisis is technological. A logical extension of this argument is that, in this particular case, the ecologist is the appropriate "engineer" to resolve the crisis. This reflects the dominant American philosophy (which is sure to come up after every lecture on the environment) that the answer to most of our problems is technology and, in particular, that the answer to the problems raised by technology is more technology. Perhaps the most astounding example of this blind faith is the recent assurance issued by the government that the SST will not fly over the United States until the sonic boom problem is solved. The sonic boom "problem," of course, cannot be "solved." One job of the ecologist is to dispel this faith in technology.

To illustrate the environmental crisis, let us take two examples of how the growth of population, combined with the increasing sophistication of technology, has caused serious problems which planning and foresight could have prevented. Unfortunately, the fact is that no technological solutions applied to problems caused by increased population have ever taken into consideration the consequences to the environment.

The first example is the building of the Aswan High Dam on the upper Nile. Its purposes were laudable—to provide a regular supply of water for irrigation, to prevent disastrous floods, and to provide electrical power for a primitive society. Other effects, however, were simply not taken into account. The annual flood of the Nile had brought a supply of rich nutrients to the eastern Mediterranean Sea, renewing its fertility; fishermen had long depended upon this annual cycle. Since the Aswan Dam put an end to the annual flood with its load of nutrients, the annual bloom of phytoplankton in the eastern Mediterranean no longer occurs. Thus the food chain from phytoplankton to zoöplankton to fish has been broken; and the sardine fishery, once producing eighteen thousand tons per year (about half of the total fish catch), has dropped to about five hundred tons per year.

Another ecological effect of the dam has been the replacement of an intermittent flowing stream with a permanent stable lake. This has allowed aquatic snails to maintain large populations, whereas before the dam was built they had been reduced each year during the dry season. Because irrigation supports larger human populations, there are now many more people living close to these stable bodies of water. The problem here is that the snails serve as intermediate hosts of the larvae of a blood fluke. The larvae leave the snail and bore into humans, infecting the liver and other organs. This causes the disease called schistosomiasis. The species of snail which lives in stable water harbors a more virulent species of fluke than that found in another species of snail in running water. Thus the lake behind the Aswan Dam has increased both the incidence and virulence of schistosomiasis among the people of the upper Nile.

A second example we might cite is the effect of DDT on the environment. DDT is only slightly soluble in water, so is carried mainly on particles in the water for short distances until these settle out. But on tiny particles in the atmosphere it is carried great distances; it may even fall out more heavily in distant places than close to where it was sprayed. DDT is not readily broken down by microörganisms; it therefore persists in the environment for many years. It is very soluble in fats so that it is quickly taken up by organisms. Herbivores eat many times their own weight of plants; the DDT is not broken down but is accumulated in their bodies and becomes further concentrated when the herbivores are eaten by the carnivores. The result is that the species at the top of the food chain end up with high doses of it in their tissues. Evidence is beginning to show that certain species of predators, such as ospreys, are being wiped out as a result of physiological debilities which lead to reproductive failure, all caused by accumulations of DDT.

The reproduction of top carnivores such as ospreys and pelicans is being reduced to negligible amounts, which will cause their extinction. No amount of technological ingenuity can reconstruct a species of osprey once it is extinct.

The tendency of DDT to kill both the herbivorous pest as well as its predators has produced some unpredicted consequences. In natural circumstances, herbivores are often kept at rather low numbers by their predators, with occasional "outbreaks" when there is a decrease in these enemies. Once spraying is started, and both the pests and their natural enemies are killed, the surviving pests, which have higher rates of increase than their predators, can then increase explosively between applications.

Before pesticides were applied to North American spruce and balsam forests, pest populations exploded once every thirty years or so, ate all the leaves, and then their numbers plummeted. Since spraying began, the pests, in the absence of a balancing force of predators, are continually able to increase between sprayings. In two instances, in cotton fields in Peru and in cocoa plantations in Malaysia, the situation became so bad that spraying was stopped. The predators returned and the damage by pests was diminished to the former tolerable levels. Another consequence of spraying has been that any member of the pest population which happens to be physiologically resistant to an insecticide survives and leaves offspring; thus resistant strains are evolved. Several hundred of these resistant strains have evolved in the last twenty years. . . .

Ecologists face problems which make their task difficult and at times apparently insurmountable. It is a young science, probably not older than forty years; consequently, much of it is still descriptive. It deals with systems which are depressingly complex, affected by dozens of variables which may all interact in a very large number of ways. Rather than taking a census of them, these systems must be sampled. Ecology is one of the few disciplines in biology in which it is not clear that removing portions of the problem to the laboratory for experimentation is an appropriate technique. It may be that the necessary simplification this involves removes exactly the elements from the system which determine how it functions. Yet field experiments are difficult to do and usually hard to interpret. Ecology, moreover, is the only field of biology which is not simply a matter of applied physics and chemistry. The great advances in molecular biology resulted from physicists looking at biological systems (such as DNA), whose basic configuration is explicable in terms of the positions of atoms. But the individual or the population is the basic unit in ecology. It seems certain, then, that a direct extension of physics and chemistry will not help ecologists.

Finally, there is the problem that each ecological situation is different from every other one, with a history all its own; ecological systems, to use a mathematical analogy, are non-Markovian, which is to say that a knowledge of both the past and the present is necessary in order to predict the future. Unlike a great deal of physics, ecology is not independent of time or place. As a consequence, the discipline does not cast up broad generalizations. All this is not a complete list of the general problems ecologists face, but it may be enough to provide a feeling for the difficulty of the subject.

Ecologists, though, do have something to show for forty years' work. These are some of the general conclusions they have reached. (Not all ecologists, by any means, would agree that they are generally applicable—and those who do agree would admit that exceptions occur—but they are the kind of basic conclusions that many ecologists would hope to be able to establish.)

Populations of most species have negative feedback processes which keep their numbers within relatively narrow limits. If the species itself does not possess such features, or even if it does, the community in which it exists acts to regulate numbers, for example, through the action of predators. (Such a statement obviously is not precise, e.g. how narrow are "relatively narrow limits"? A measure of ecology's success, or lack of it, is that, in forty years, there are no more than a half-dozen populations in which regulation has been adequately demonstrated; and the basis for belief in regulation is either faith or very general observations, such as the fact that most species are not so abundant that they are considered pests.)

The laws of physics lead to derivative statements in ecology. For example, the law that matter cycles through the ecosystem, to be used again and again. Or the law that energy from the sun is trapped by plants through photosynthesis, moves up the food chain to herbivores and then to carnivores as matter, losing energy at each successive conversion so that there is generally less energy and biomass in higher food levels than in lower ones. Ecologists have tried to take such truths from physics and construct more truly ecological generalities from them. Thus, to stay with the same example, it appears likely that there are never more than five links in any one chain of conversions from plant to top predator.

It is probably true, on a given piece of the earth and provided that the climate doesn't change, that a "climax" ecosystem will develop which is characteristic of the area's particular features and that places with similar features will develop similar ecosystems if left undisturbed. Characteristically, a "succession" from rather simple and short-lived communities to more complex and more persistent communities will occur, though there may be a reduction in the complexity of the final community. We use "final" to mean that a characteristic community will be found there for many generations. We might go further and say that during the period of development disturbances of the community will result in its complexity being reduced. (Again, such statements will certainly arouse the dissent of some ecologists.)

Finally, most ecologists would agree that complex communities are more stable than simple communities. This statement illustrates the difficulties faced by theoretical ecologists. Take some of its implications: What is complexity and what is stability in an ecological setting? Charles Elton embodied the idea in a simple, practical, and easily understood way. He argued that England should maintain the hedgerows between its fields because these were complex islands in a simple agricultural sea and contained a reservoir of insect and other predators which helped to keep down pest populations. The idea here seems quite clear. Ecologists, though, want a more precise exposition of the implications of the statement. What kind of complexity? What is stability?

Physical complexity, by providing hiding places for prey, may increase stability. Certainly biological complexity in general is thought to lead to stability—more species or more interspecific interactions, more stability. But we may ask, more species of what sort? Here a variety of answers is available. It has been suggested that complex communities are stable, i.e. able to resist invasion by species new to the area, by having all the "niches" filled. Thus sheer numbers of kinds of organisms in all food levels were considered the appropriate sort of complexity. To keep the numbers of prey stable, the most likely candidates are predators. Now other questions arise: Do we just want more species of predators? Do we want more species of predators which are very specific in the prey they eat, implying that prey are stabilized by having many species feed on them? Do we want predators which are very general and attack many prey species, so that we still have a large number of interspecific interactions which are made up in a different way? The answer is not obvious, and indeed there is disagreement on it. Furthermore, if one studies the way some predators react to changes in the numbers of their prey, their short-term responses are such as to cause *instability*. Thus only some types of biological complexity may produce stability.

What do we mean by stability? In the examples cited, we have meant numerical constancy through time, but this is by no means the only meaning. It has even been suggested that numerical *in*constancy is a criterion for stability. Stability might also mean that the same species persist in the same area over long periods, showing the same sort of interspecific interactions (community stability). A community or population might be considered stable because it does not change in response to a great deal of environmental pressure, or because it changes but quickly returns to its original state when the disturbing force is removed. It is worth noting that if a population or community is observed merely not to change, we cannot tell whether this is owing to its ability to resist perturbing factors or merely to the absence of such factors. If we want to know about the *mechanisms* which might lead to the truth of our original statement, "complexity leads to stability," all the above points are important.

This general statement about complexity and stability rests upon the kind of observation readily apparent to most intelligent laymen. Thus simple agricultural systems seem to be much more subject to outbreaks of herbivores than the surrounding countryside. Ecosystems in the tropics appear to be more stable than in the simpler temperate zone. In turn the temperate zone seems to be more stable than the Arctic. This seems to be mainly an article of faith. However, even this classic sort of evidence is questioned—for example, small mammals may actually be more unstable numerically in the United States than in the much simpler Arctic environment. Other evidence comes from the laboratory. If one takes small species of prey and predator—for example, two single-celled animals or two small mites—and begins culturing them together, the numbers of prey and predators fluctuate wildly and then both become extinct quickly, for the predators exhaust their food source. "Simple" predator-prey systems tend to be unstable. There is some evidence that if physical complexity is added the system may become more stable.

From these examples of the generalizations ecologists have arrived at, an important question emerges. Even if we dispense with the idea that ecologists are some sort of environmental engineers and compare them to the pure physicists who provide scientific rules for engineers, do the tentative understandings we have outlined provide a sound basis for action by those who would manage the environment? It is self-evident that they do not.

This conclusion seems to be implied in a quotation from an article published in *Time* on the environment, which underlines the point that application of the ecologist's work is not the solution to the environmental crisis. According to *Time:* "Crawford S. Holling was once immersed in rather abstract research at the University of British Columbia—mathematical models of the relationship between predators and their prey. 'Three years ago, I got stark terrified at what was going on in the world and gave it up.' Now he heads the university's interdepartmental studies of land and water use, which involve agriculture, economics, forestry, geography, and regional planning. 'What got me started on this,' says Holling, 'were the profound and striking similarities between ecological systems and the activities of man: between predators and land speculators; between animal-population growth and economic growth; between plant dispersal and the diffusion of people, ideas, and money.' "

The "rather abstract research" was ecology. Holling's testimony is that it would not provide a solution. Yet, by and large, ecologists are concerned and probably have the best understanding of the problem.

We submit that ecology as such probably cannot do what many people expect it to do; it cannot provide a set of "rules" of the kind needed to manage the environment. Nevertheless, ecologists have a great responsibility to help solve the crisis; the solution they offer should be founded on a basic "ecological attitude." Ecologists are likely to be aware of the consequences of environmental manipulation; possibly most important, they are ready to deal with the environmental problem since their basic ecological attitude is itself the solution to the problem. Interestingly enough, the supporting data do not generally come from our "abstract research" but from massive uncontrolled "experiments" done in the name of development.

These attitudes and data, plus obvious manifestations of physical laws, determine what the ecologist has to say on the problem and constitute what might be called environmental knowledge. Some examples of this knowledge follow, though this is not to be taken as an encapsulation of the ecologist's wisdom.

Whatever is done to the environment is likely to have repercussions in other places and at other times. Because of the characteristic problems of ecology some of the effects are bound to be unpredictable in practice, if not in principle. Furthermore, because of the characteristic time-dependence problem, the effects may not be measurable for years—possibly not for decades.

If man's actions are massive enough, drastic enough, or of the right sort, they will cause changes which are irreversible since the genetic material of extinct species cannot be reconstituted. Even if species are not driven to extinction, changes may occur in the ecosystem which prevent a recurrence of the events which produced the community. Such irreversible changes will almost always produce a simplification of the environment.

The environment is finite and our non-renewable resources are finite. When the stocks run out we will have to recycle what we have used.

The capacity of the environment to act as a sink for our total waste, to absorb it and recycle it so that it does not accumulate as pollution, is limited. In many instances, that limit has already been passed. It seems clear that when limits are passed, fairly gross effects occur, some of which are predictable, some of which are not. These effects result in significant alterations in environmental conditions (global weather, ocean productivity). Such changes are almost always bad since organisms have evolved and ecosystems have developed for existing conditions. We impose rates of change on the environment which are too great for biological systems to cope with.

In such a finite world and under present conditions, an increasing population can only worsen matters. For a stationary population, an increase in standard of living can only mean an increase in the use of limited resources, the destruction of the environment, and the choking of the environmental sinks.

There are two ways of attacking the environmental crisis. The first approach is technology; the second is to reverse the trends which got us into the crisis in the first place and to alter the structure of our society so that an equilibrium between human population and the capacities of the environment can be established.

There are three main dangers in a technological approach to the environmental crisis. The first threatens the environment in the short term, the second concerns ecologists themselves, and the third, which concerns the general public attitude, is a threat to the environment in the long term.

Our basic premise is that, by its nature, technology is a system for manufacturing the need for more technology. When this is combined with an economic system whose major goal is growth, the result is a society in which conspicuous production of garbage is the highest social virtue. If our premise is correct, it is unlikely we can solve our present problems by using technology. As an example, we might consider nuclear power plants as a "clean" alternative to which we can increasingly turn. But nuclear power plants inevitably produce radioactive waste; this problem will grow at an enormous rate, and we are not competent to handle it safely. In addition, a whole new set of problems arises when all these plants produce thermal pollution. Technology merely substitutes one sort of pollution for another.

There is a more subtle danger inherent in the technological approach. The automobile is a blight on Southern California's landscape. It might be thought that ecologists should concern themselves with encouraging the development of technology to cut down the emission of pollutants from the internal combustion engine. Yet that might only serve to give the public the impression that something is being done about the problem and that it can therefore confidently await its solution. Nothing significant could be accomplished in any case because the increasing number of cars ensures an undiminishing smog problem.

Tinkering with technology is essentially equivalent to oiling its wheels. The very act of making minor alterations, in order to placate the public, actually allows the general development of technology to proceed unhindered, only increasing the environmental problems it causes. This is what sociologists have called a "pseudo-event." That is, activities go on which give the appearance of tackling the problem; they will not, of course, solve it but only remove public pressure for a solution.

Tinkering also distracts the ecologist from his real job. It is the ecologist's job, as a general rule, to oppose growth and "progress." He cannot set about convincing the public of the correctness of this position if in the meantime he is putting his shoulder behind the wheel of technology. The political power system has a long tradition of buying off its critics, and the ecologist is liable to wind up perennially compromising his position, thereby merely slowing down slightly or redirecting the onslaught of technology.

The pressures on the ecologist to provide "tinkering" solutions will continue to be quite strong. Pleas for a change of values, for a change to a non-growth, equilibrium economy seem naive. The government, expecting sophistication from its "experts," will probably receive such advice coolly. Furthermore, ecologists themselves are painfully aware of how immature their science is and generally take every opportunity to cover up this fact with a cloud of obfuscating pseudo-sophistication. They delight in turning prosaic facts and ideas into esoteric jargon. Where possible, they embroider the structure with mathematics and the language of cybernetics and systems analysis, which is sometimes useful but frequently is merely confusing. Such sophistication is easily come by in suggesting technological solutions.

Finally, there is always the danger that in becoming a governmental consultant, the ecologist will aim his sights at the wrong target. The history of the Washington "expert" is that he is called in to make alterations in the model already decided upon by the policymakers. It would be interesting to know what proportion of scientific advice has ever produced a change in ends rather than in means. We suspect it is minute. But the ecologist ought not to concern himself with less than such a change; he must change the model itself.

We should point out that we are not, for example, against substituting a steam-driven car for a gas-driven car. Our contention is that by changing public attitudes the ecologist can do something much more fundamental. In addition, by changing these attitudes he may even make it easier to force the introduction of "cleaner" technology, since this also is largely a *political* decision. This certainly seems to be so in the example of the steam-driven car.

We do not believe that the ecologist has anything really new to say. His task, rather, is to inculcate in the government and the people basic ecological attitudes. The population must come, and very soon, to appreciate certain basic notions. For example: a finite world cannot support or withstand a continually expanding population and technology; there are limits to the capacity of environmental sinks; ecosystems are sets of interacting entities and there is no "treatment" which does not have "side effects" (e.g. the Aswan Dam); we cannot continually simplify systems and expect them to remain stable, and once they do become unstable there is a tendency for instability to increase with time. Each child should grow up knowing and understanding his place in the environment and the possible consequences of his interaction with it.

In short, the ecologist must convince the population that the only solution to the problem of growth is not to grow. This applies to population and, unless the population is declining, to its standard of living. It should be clear by now that "standard of living" is probably beginning to have an inverse relationship to the quality of life. An increase in the gross national product must be construed, from the ecological point of view, as disastrous. (The case of underdeveloped countries, of course, is different.)

We do not minimize the difficulties in changing the main driving force in life. The point of view of the ecologist, however, should be subversive; it has to be subversive or the ecologist will become merely subservient. Such a change in values and structure will have profound consequences. For example, economists, with a few notable exceptions, do not seem to have given any thought to the possibility or desirability of a stationary economy. Businessmen, and most economists, think that growth is good, stagnation or regression is bad. Can an equilibrium be set up with the environment in a system having this philosophy? The problem of converting to non-growth is present in socialist countries too, of course, but we must ask if corporate capitalism, by its nature, can accommodate such a change and still retain its major features. By contrast, if there are any ecological laws at all, we believe the ecologists' notion of the inevitability of an equilibrium between man and the environment is such a law. . . .

Ian L. McHarg: Values, Process, and Form

It is my proposition that, to all practical purposes, western man remains obdurately pre-Copernican, believing that he bestrides the earth round which the sun, the galaxy, and the very cosmos revolve. This delusion has fueled our ignorance in time past and is directly responsible for the prodigal destruction of nature and for the encapsulating burrows that are the dysgenic city.

We must see nature and man as an evolutionary process which responds to laws, which exhibits direction, and which is subject to the final test of survival. We must learn that nature includes an intrinsic value-system in which the currency is energy and the inventory is matter and its cycles—the oceans and the hydrologic cycle, life-forms and their roles, the cooperative mechanisms which life has developed and, not least, their genetic potential. . . .

This can be pictured simply in a comparison between the early earth and the present planet. In the intervening billions of years the earth has been transformed and the major change has been in the increase of order. Think of the turbulence and violence of the early earth, racked by earthquakes and vulcanism, as it evolved toward equilibrium, and of the unrestrained movements of water, the dust storms of unstabilized soils, and the extreme alternations of climate unmodified by a green, meliorative vegetative cover. In this early world physical processes operated toward repose, but in the shallow bays there emerged life and a new kind of ordering was initiated. The atmosphere which could sustain life was not the least of the creations of life. Life elaborated in the seas and then colonized the earth, thus increasing the opportunities for life and for evolution. Plants and decomposers created the soils, anchored the surface of the earth, checked the movements of soil particles, modified water processes, meliorated the climate, and ordered the distribution of nutrients. Species evolved to occupy and create more habitats, more niches, each increase requiring new cooperative relationships between organisms—new roles, all of which were beneficial. In the earth's history can be seen the orderings which life has accomplished: the increase to life forms, habitats and roles, symbiotic relationships, and the dynamic equilibrium in the system—the total an increase in order. This is creation.

From *The Fitness of Man's Environment,* Smithsonian Institution Press, 1968. Reprinted by permission of the publisher.

In the early earth, the sunlight which fell upon the planet equaled the degraded energy which was radiated from it. Since the beginning of plant-life, some of the sun's energy has been entrapped by photosynthesis and employed with matter to constitute the ordered beings of plants; thence, to the animals and decomposers, and all of the orderings which they have accomplished. This energy will surely be degraded, but the entrapped energy, with matter, is manifest in all life forms past and present, and in all of the orderings which they have accomplished. Thus, creation equals the energy which has been temporarily entrapped and used with matter to accomplish all of the ordering of physical, biological, and cultural evolution. This, physicists describe as negentropy, in contrast with the inevitable degradation of energy which is described as entropy.

By this we see the world as a creative process involving all matter and all life forms in all time past and in the present. Thus, creation reveals two forms: first, the physical entrapment and ordering which is accomplished primarily by plants and by the simplest animals; and, second, apperception and the resulting ordering for which an increased capacity is observed as species rise in the phylogenetic scale. In this, man is seen to be especially endowed. This view of the world as a creative process involving all of its denizens, including man, in a cooperative enterprise, is foreign to the western tradition that insists upon the exclusive divinity of man, his independent superiority, dominion, and license to subjugate the earth. It is this man in whose image was God made. This concept of nature as a creative, interacting process in which man is involved with all other life forms is the ecological view. It is, I submit, the best approximation of the world that has been presented to us, and the indispensable approach to determining the role of man in the biosphere. It is indispensable also for investigation, not only of the adaptations which man accomplishes, but of their forms.

The place, the plants, the animals, and man, and the orderings which they have accomplished over time, are revealed in form. To understand this it is necessary to invoke all physical, biological, and cultural evolution. Form and process are indivisible aspects of a single phenomenon: being. Norbert Weiner described the world as consisting of "To Whom It May Concern" messages, but these are clothed in form. Process and fitness (which is the criterion of process) are revealed in form; form contains meaning. The artifact, tool, room, street, building, town or city, garden or region, can be examined in terms of process, manifest in form, which may be unfit, fit, or most fitting. The last of these, when made by man, is art.

The role of man is to understand nature, which is also to say man, and to intervene to enhance its creative processes. He is the prospective steward of the biosphere. The fruits of the anthropocentric view are in the improvement of the social environment, and great indeed are their values, but an encomium on social evolution is not my competence, and I leave the subject with the observation that, while Madison, Jefferson, Hamilton, and

Washington might well take pride in many of our institutions, it is likely that they would recoil in horror from the face of the land of the free.

An indictment of the physical environment is too easy, for post-industrial cities are such squalid testimony to the bondage of toil and to the insensitivity of man, that the most casual examination of history reveals the modern city as a travesty of its antecedents and a denial of its role as the proudest testimony to peoples and their cultures. The city is no longer the preferred residence for the polite, the civilized, and the urbane, all of which say "city." They have fled to the illusion of the suburb, escaping the iridescent shills, neon vulgarity of the merchants, usurious slumlords, cynical polluters (household names for great corporations, not yet house-broken), crime, violence, and corruption. Thus, the city is the home of the poor, who are chained to it, and the repository of dirty industry and the commuter's automobile. Give us your poor and oppressed, and we will give them Harlem and the Lower East Side, Bedford-Stuyvesant, the South Side of Chicago, and the North of Philadelphia—or, if they are very lucky, Levittown. Look at one of these habitats through the Cornell Medical School study of Midtown Manhattan, where 20 percent of a sample population was found to be indistinguishable from the patients in mental institutions, and where a further 60 percent evidenced mental disease. Observe the environments of physical, mental, and social pathology. What of the countryside? Well, you may drive from the city and search for the rural landscape, but to do so you will follow the paths of those who preceded you, and many of them stayed to build. But those who did so *first* are now deeply embedded in the fabric of the city. So as you go you will transect the annular rings of the thwarted and disillusioned who are encapsulated in the city as nature endlessly eludes pursuit. You can tell when you have reached the edge of the rural scene for there are many emblems: the cadavers of old trees, piled in untidy heaps beside the magnificent machines for land despoliation, at the edge of the razed deserts; forests felled; marshes filled; farms obliterated; streams culverted; and the sweet rural scene transformed into the ticky-tacky vulgarity of the merchants' creed and expression. . . .

This is the face of our western inheritance—Judaism, Christianity, Humanism, and the Materialism which is better named Economic Determinism. The countryside, the last great cornucopia of the world's bounty, ravaged; and the city of man (God's Junkyard, or call it Bedlam) a vast demonstration of man's inhumanity to man, where existence, sustained by modern medicine and social legislation, is possible in spite of the physical environment. Yet we are the inheritors of enormous beauty, wealth, and variety. . . . It is necessary to resolve to fulfill the American Revolution and to create the fair image that can be the land of the free and the home of the brave. But to resolve is not enough; it is also necessary that society at large understand nature as a process, having values, limiting factors, opportunities, and constraints; that creation and destruction are real; that there are criteria by which we can discern the direction and tests of evolution; and, finally, that there are formal implications revealed in the environment which affect the nature and form of human adaptations.

What inherited values have produced this plight, from which we must be released if the revolution is to be completed? Surely it is the very core of our tradition, the Judeo-Christian-Humanist view which is so unknowing of nature and of man, which has bred and sustained this simple-minded anthropocentrism and anthropomorphism. It is this obsolete view of man and nature which is the greatest impediment to our emancipation as managers of the countryside, city builders, and artists. If it requires little effort to mobilize a sweeping indictment of the physical environment which is man's creation, it takes little more to identify the source of the value system which is the culprit. Whatever the origins, the text is quite clear in Judaism, was absorbed all but unchanged into Christianity, and was inflated in Humanism to become the implicit attitude of western man to nature and the environment. Man is exclusively divine, all other creatures and things occupy lower and generally inconsequential status; man is given dominion over all creatures and things; he is enjoined to subdue the earth. Here is the best of all possible texts for him who would contemplate biocide, carelessly extirpate great realms of life, create Panama Canals, or dig Alaskan harbors with atomic demolition. Here is the appropriate injunction for the land rapist, the befouler of air and water, the uglifier, and the gratified bulldozer. Dominion and subjugation, or better call it conquest, are their creeds. . . .

The face of the city and the land are the best testimony to the concept of conquest and exploitation—the merchants' creed. The Gross National Product is the proof of its success, money is its measure, convenience is its cohort, the short term is its span, and the devil take the hindmost is its morality. The economists, with some conspicuous exceptions, have become the spokesmen for the merchants' creed and in concert they ask with the most barefaced affrontery that we accommodate our values to theirs. Neither love nor compassion, health nor beauty, dignity nor freedom, grace nor delight are true unless they can be priced. If not, they are described as nonprice benefits and relegated to inconsequence, and the economic model proceeds towards its self-fulfillment—which is to say more despoliation. The major criticism of this model is not that it is partial (which is conceded by its strongest advocates), but more that the features which are excluded are among the most important human values, and also the requirements for survival. If the ethics of society insist that it is man's bounden duty to subdue the earth, then it is likely that he will obtain the tools with which to accomplish this. If there is established a value system based upon exploitation of the earth, then the essential components for survival, health, and evolution are likely to be discounted, as they are. It can then come as no surprise to us that the most scabrous slum is more highly valued than the most beautiful landscape, that the most loathsome roadside stand is more highly valued than the richest farmland, and that this society should more highly prize tomato stakes than the primeval redwoods whence they come.

It is, in part, understandable why our economic value system is completely blind to the realities of the biophysical world—why it excludes from consideration, not only the most important human aspirations, but even those processes which are indispensable for survival. The origins of society and exchange began in an early world where man was a trifling inconsequence in the face of an overwhelming nature. He knew little of its operation. He bartered his surpluses of food and hides, cattle, sheep and goats; and valued such scarcities as gold, silver, myrrh and frankincense. In the intervening millennia the valuations attributed to commodities have increased in range and precision and the understanding of the operation of this limited sphere has increased dramatically. Yet, we are still unable to identify and evaluate the processes which are indispensable for survival. When you give money to a broker to invest you do so on the understanding that this man understands a process well enough to make the investment a productive one. Who are the men to whom you entrust the responsibility for ensuring a productive return on the world's investment? Surely, those who understand physical and biological processes, realize that these are creative. The man who views plants as the basis of negentropy in the world and the base of the food chain, as the source of atmospheric oxygen, fossil fuels and fibers, is a different man from one who values only economic plants, or that man who considers them as decorative but irrelevant aspects of life. The man who sees the sun as the source of life and the hydrologic cycle as its greatest work, is a different man from one who values sunlight in terms of a recreation industry, a portion of agricultural income, or from that man who can obscure sky and sunlight with air pollution, or who carelessly befouls water. The man who knows that the great recycling of matter, the return stroke in the world's cycles, is performed by the decomposer bacteria, views soils and water differently from the man who values a few bacteria in antibiotics, or he who is so unknowing of bacteria that he can blithely sterilize soils and make streams septic. That man who has no sense of the time which it has taken for the elaboration of life and symbiotic arrangements which have evolved, can carelessly extirpate creatures. That man who knows nothing of the value of the genetic pool, the greatest resource which we bring to the future, is not likely to fear radiation hazard or to value life. Clearly, it is illusory to expect the formulation of a precise value system which can include the relative value of sun, moon, stars, the changing seasons, physical processes, life forms, their roles, their symbiotic relationships, or the genetic pool. Yet, without precise evaluation, it is apparent that there will be a profound difference in attitude—indeed, a profoundly different value system—between those who understand the history of evolution and the interacting processes of the biosphere, and those who do not.

. . . We need a general theory which encompasses physical, biological, and cultural evolution; which contains an intrinsic value system; which includes criteria of creativity and destruction and, not least, principles by which we can measure adaptations and their form. Surely, the minimum requirement for an attitude to nature and to man is that it approximate reality. Clearly, our traditional view does not. If one would know of these things, where else should one turn but to science. If one wishes to know of the phenomenal world, where better to ask than the natural sciences; if you would know of the interactions between organism and environment, then turn to the ecologist, for this is his competence. From the ecological view, one can conclude that by living one is united physically to the origins of life. If life originated from matter, then by living one is united with the primeval hydrogen. The earth has been the one home for all of its evolving processes and for all of its inhabitants; from hydrogen to man, it is only the bathing sunlight which changes. The planet contains our origins, our history, our milieu—it is our home. It is in this sense that ecology, derived from oikos, is the science of the home. Can we review physical and biological evolution to discern the character of these processes, their direction, the laws which obtain, the criteria for survival and success? If this can be done, there will also be revealed an intrinsic value system and the basis for form. This is the essential ingredient of an adequate view of the world: a value system which corresponds to the creative processes of the world, and both a diagnostic and constructive view of human adaptations and their form.

. . . Each creature must adapt to the others in that concession of autonomy toward the end of survival that is symbiosis. Thus parasite and host, predator and prey, and those creatures of mutual benefit develop symbioses to ensure survival. The world works through cooperative mechanisms in which the autonomy of the individual, be it cell, organ, organism, species, or community is qualified toward the survival and evolution of higher levels of order culminating in the biosphere. Now these symbiotic relationships are beneficial to the sum of organisms although clearly many of them are detrimental to individuals and species. While the prey is not pleased with the predator or the host far from enamored of the parasite or the pathogen, these are regulators of populations and the agents of death—that essential return phase in the cycle of matter, which fuels new life and evolution. Only in this sense can the predator, parasite, and pathogen be seen as important symbiotic agents, essential to the creative processes of life and evolution. If evolution has proceeded from simple to complex, this was accomplished through symbiosis. As the number of species increased, so then did the number of roles and the symbiotic arrangements between species. If stability increases as evolution proceeds, then this is the proof of increased symbiosis. . . .

What conclusions can one reach from this investigation? The first is that the greatest failure of western society, and of the post-industrial period in particular, is the despoliation of the natural world and the inhibition of life which is represented by modern cities. It is apparent that this is the inevitable consequence of the values that have been our inheritance. It is clear, to me if to no one else, that these values have little correspondence to reality and perpetrate an enormous delusion as to the world, its work, the importance of the roles that are performed, and, not least, the potential role of man. In this delusion the economic model is conspicuously inadequate, excluding as it does the most important human aspirations and the realities of the biophysical world. The remedy requires that the understanding of this world which now reposes in the natural sciences be absorbed into the conscious value system of society, and that we learn of the evolutionary past and the roles played by physical processes and life forms. We must learn of the criteria for creation and destruction, and of the attributes of both. We need to formulate an encompassing value system which corresponds to reality and which uses the absolute values of energy, matter, life forms, cycles, roles, and symbioses.

. . . We must abandon the self mutilation which has been our way, reject the title of planetary disease which is so richly deserved, and abandon the value system of our inheritance which has so grossly misled us. We must see nature as process within which man exists, splendidly equipped to become the manager of the biosphere; and give form to that symbiosis which is his greatest role, man the world's steward.

Aldo Leopold: The Conservation Ethic

When godlike Odysseus returned from the wars in Troy, he hanged all on one rope some dozen slave-girls of his household whom he suspected of misbehavior during his absence.

This hanging involved no question of propriety, much less of justice. The girls were property. The disposal of property was then, as now, a matter of expediency, not of right and wrong.

Criteria of right and wrong were not lacking from Odysseus' Greece: witness the fidelity of his wife through the long years before at last his black-prowed galleys clove the wine-dark seas for home. The ethical

From the *Journal of Forestry,* October 1933. Reprinted by permission of The Society of American Foresters.

structure of that day covered wives, but had not yet been extended to human chattels. During the three thousand years which have since elapsed, ethical criteria have been extended to many fields of conduct, with corresponding shrinkages in those judged by expediency only.

This extension of ethics, so far studied only by philosophers, is actually a process in ecological evolution. Its sequences may be described in biological as well as philosophical terms. An ethic, biologically, is a limitation on freedom of action in the struggle for existence. An ethic, philosophically, is a differentiation of social from antisocial conduct. These are two definitions of one thing. The thing has its origins in the tendency of interdependent individuals or societies to evolve modes of cooperation. The biologist calls these symbioses. Man elaborated certain advanced symbioses called politics and economics. Like their simpler biological antecedents, they enable individuals or groups to exploit each other in an orderly way. Their first yardstick was expediency.

The complexity of coöperative mechanisms increased with population density, and with the efficiency of tools. It was simpler, for example, to define the antisocial uses of sticks and stones in the days of the mastodons than of bullets and billboards in the age of motors.

At a certain stage of complexity, the human community found expediency yardsticks no longer sufficient. One by one it has evolved and superimposed upon them a set of ethical yardsticks. The first ethics dealt with the relationship between individuals. The Mosaic Decalogue is an example. Later accretions dealt with the relationship between the individual and society. Christianity tries to integrate the individual to society, democracy to integrate social organization to the individual.

There is as yet no ethic dealing with man's relationship to land and to the nonhuman animals and plants which grow upon it. Land, like Odysseus' slave-girls, is still property. The land relation is still strictly economic, entailing privileges but not obligations.

The extension of ethics to this third element in human environment is, if we read evolution correctly, an ecological possibility. It is the third step in a sequence. The first two have already been taken. Civilized man exhibits in his own mind evidence that the third is needed. For example, his sense of right and wrong may be aroused quite as strongly by the desecration of a nearby woodlot as by a famine in China, a near-pogrom in Germany, or the murder of the slave-girls in ancient Greece. Individual thinkers since the days of Ezekial and Isaiah have asserted that the despoliation of land is not only inexpedient but wrong. Society, however, has not yet affirmed their belief. I regard the present conservation movement as the embryo of such an affirmation. I here discuss why this is, or should be, so.

Some scientists will dismiss this matter forthwith, on the ground that ecology has no relation to right and wrong. To such I reply that science, if not philosophy, should by now have made us cautious about dismissals. An ethic may be regarded as a mode of guidance for meeting ecological situations so new or intricate, or involving such deferred reactions, that the path of social expediency is not discernible to the average individual. Animal instincts are just this. Ethics are possibly a kind of advanced social instinct in the making.

Whatever the merits of this analogy, no ecologist can deny that our land relation involves penalties and rewards which the individual does not see, and needs modes of guidance which do not yet exist. Call these what you will, science cannot escape its part in forming them. . . .

The Economic Isms

As nearly as I can see, all the new isms—Socialism, Communism, Fascism, and especially the late but not lamented Technocracy—outdo even Capitalism itself in their preoccupation with one thing: The distribution of more machine-made commodities to more people. They all proceed on the theory that if we can all keep warm and full, and all own a Ford and a radio, the good life will follow. Their programs differ only in ways to mobilize machines to this end. Though they despise each other, they are all, in respect of this objective, as identically alike as peas in a pod. They are competitive apostles of a single creed: *salvation by machinery.*

We are here concerned, not with their proposals for adjusting men and machinery to goods, but rather with their lack of any vital proposal for adjusting men and machines to land. To conservationists they offer only the old familiar palliatives: Public ownership and private compulsion. If these are insufficient now, by what magic are they to become sufficient after we change our collective label?

Let us apply economic reasoning to a sample problem and see where it takes us. As already pointed out, there is a huge area which the economist calls submarginal, because it has a minus value for exploitation. In its once-virgin condition, however, it could be "skinned" at a profit. It has been, and as a result erosion is washing it away. What shall we do about it?

By all the accepted tenets of current economics and science we ought to say "let her wash." Why? Because staple land crops are overproduced, our population curve is flattening out, science is still raising the yields from better lands, we are spending millions from the public treasury to retire unneeded acreage, and here is nature offering to do the same thing free of charge; why not let her do it? This, I say, is economic reasoning. *Yet no man has so spoken.* I cannot help reading a meaning into this fact. To me it means that the average citizen shares in some degree the intuitive and instantaneous contempt with which the conservationist would regard such an attitude. We can, it seems, stomach the burning or plowing-under of overproduced cotton, coffee, or corn, but the destruction of mother earth, however "submarginal," touches something deeper, some subeconomic stratum of the human intelligence wherein lies that something—perhaps the essence of civilization—which Wilson called "the decent opinion of mankind."

The Conservation Movement

We are confronted, then, by a contradiction. To build a better motor we tap the uttermost powers of the human brain; to build a better countryside we throw dice. Political systems take no cognizance of this disparity, offer no sufficient remedy. There is, however, a dormant but widespread consciousness that the destruction of land, and of the living things upon it, is wrong. A new minority have espoused an idea called conservation which tends to assert this as a positive principle. Does it contain seeds which are likely to grow?

Its own devotees, I confess, often give apparent grounds for skepticism. We have, as an extreme example, the cult of the barbless hook, which acquires self-esteem by a self-imposed limitation of armaments in catching fish. The limitation is commendable, but the illusion that it has something to do with salvation is as naive as some of the primitive taboos and mortifications which still adhere to religious sects. Such excrescences seem to indicate the whereabouts of a moral problem, however irrelevant they be in either defining or solving it.

Then there is the conservation-booster, who of late has been rewriting the conservation ticket in terms of "tourist bait." He exhorts us to "conserve outdoor Wisconsin" because if we don't the motorist on vacation will streak through to Michigan, leaving us only a cloud of dust. Is Mr. Babbitt trumping up hard-boiled reasons to serve as a screen for doing what he thinks is right? His tenacity suggests that he is after something more than tourists. Have he and other thousands of "conservation workers" labored through all these barren decades fired by a dream of augmenting the sales of sandwiches and gasoline? I think not. Some of these people have hitched their wagon to a star—and that is something.

Any wagon so hitched offers the discerning politician a quick ride to glory. His agility in hopping up and seizing the reins adds little dignity to the cause, but it does add the testimony of his political nose to an important question: is this conservation something people really want? The political objective, to be sure, is often some trivial tinkering with the laws, some useless appropriation, or some pasting of pretty labels on ugly realities. How often, though, does any political action portray the real depth of the idea behind it? For political consumption a new thought must always be reduced to a posture or a phrase. It has happened before that great ideas were heralded by growing pains in the body politic, semicomic to those onlookers not yet infected by them. The insignificance of what we conservationists, in our political capacity, say and do, does not detract from the significance of our persistent desire to do something. To turn this desire into productive channels is the task of time, and ecology.

. . . Granted a community in which the combined beauty and utility of land determines the social status of its owner, and we will see a speedy dissolution of the economic obstacles which now beset conservation. Economic laws may be permanent, but their impact reflects what people want, which in turn reflects what they know and what they are. The economic setup at any moment is in some measure the result, as well as the cause, of the then prevailing standard of living. Such standards change. For example: some people discriminate against manufactured goods produced by child labor or other antisocial processes. They have learned some of the abuses of machinery, and are willing to use their custom as a leverage for betterment. Social pressures have also been exerted to modify ecological processes which happened to be simple enough for people to understand—witness the very effective boycott of birdskins for millinery ornament. We need postulate only a little further advance in ecological education to visualize the application of like pressures to other conservation problems.

For example: the lumberman who is now unable to practice forestry because the public is turning to synthetic boards may be able to sell man-grown lumber "to keep the mountains green." Again: certain wools are produced by gutting the public domain; couldn't their competitors, who lead their sheep in greener pastures, so label their product? Must we view forever the irony of educating our sons with paper, the offal of which pollutes the rivers which they need quite as badly as books? Would not many people pay an extra penny for a "clean" newspaper? Government may some day busy itself with the legitimacy of labels used by land industries to distinguish conservation products, rather than with the attempt to operate their lands for them.

I neither predict nor advocate these particular pressures—their wisdom or unwisdom is beyond my knowledge. I do assert that these abuses are just as real, and their correction every whit as urgent, as was the killing of egrets for hats. *They differ only in the number of links composing the ecological chain of cause and effect.* In egrets there were one or two links, which the mass-mind saw, believed, and acted upon. In these others there are many links; people do not see them, nor believe us who do. The ultimate issue, in conservation as in other social problems, is whether the mass-mind *wants* to extend its powers of comprehending the world in which it lives, or, granted the desire, *has the capacity to do so*. Ortega, in his "Revolt of the Masses," has pointed the first question with devastating lucidity. The geneticists are gradually, with trepidations, coming to grips with the second. I do not know the answer to either. I simply affirm that a sufficiently enlightened society, by changing its wants and tolerances, can change the economic factors bearing on land. It can be said of nations, as of individuals: "as a man thinketh, so is he."

It may seem idle to project such imaginary elaborations of culture at a time when millions lack even the means of physical existence. Some may feel for it the same honest horror as the Senator from Michigan who lately arraigned Congress for protecting migratory birds at a time when fellow humans lacked bread. The trouble with such deadly parallels is we can never be sure which is cause and which is effect. It is not inconceivable that the wave phenomena which have lately upset everything from banks to crime rates might be less troublesome if the human medium in which they run *readjusted its tensions.* The stampede is an attribute of animals interested solely in grass.

Garrett de Bell: Energy

All power pollutes.

Each of the major forms of power generation does its own kind of harm to the environment. Fossil fuels—coal and oil—produce smoke and sulfur dioxide at worst; even under ideal conditions they convert oxygen to carbon dioxide. Hydroelectric power requires dams that cover up land, spoil wild rivers, increase water loss by evaporation, and eventually produce valleys full of silt. Nuclear power plants produce thermal and radioactive pollution and introduce the probability of disaster.

We are often told that it is essential to increase the amount of energy we use in order to meet demand. This "demand," we are told, must be met in order to increase or maintain our "standard of living." What these statements mean is that if population continues to increase, and if per-capita power continues to increase as in the past, then power generation facilities must be increased indefinitely.

Such statements ignore the environmental consequences of building more and more power generation facilities. They ignore the destruction of wild rivers by dams, the air pollution by power plants, the increasing danger of disease and disaster from nuclear power facilities.

These effects can no longer be ignored, but must be directly con-

---

fronted. *The perpetually accelerating expansion of power output is not necessary.*

It is assumed by the utilities that the demand for power is real because people continue to purchase it. However, we are all bombarded with massive amounts of advertising encouraging us to buy appliances, gadgets, new cars, and so on. There is no comparable public service advertising pointing up the harmful effects of over-purchase of "convenience" appliances that increase use of power. Public utilities aggressively advertise to encourage increasing use of power. For instance, Pacific Gas and Electric advertises: "Beautify America—use electric clothes dryers." The unbeautifying results of building more power plants are, of course, not mentioned.

For the lopsided advertising, public utilities use public monies, paid in by the consumer. This is allowed by the regulatory agencies on the theory that increasing use of power lowers the per unit cost, which is beneficial to the consumer. However, the consumer is also the person who breathes the polluted air and has his view spoiled by a power plant. Therefore, this sort of advertising should be prohibited.

But perhaps it is unrealistic to expect the power companies and the appliance and car builders to call a halt, to flatly say, "This is where we stop. The limits have been reached, even exceeded." The limits can, and must, be set by the consumer. It is the consumer, ultimately, who must decide for himself what appliances he needs and which he can forego. The producers of power and power-using appliances will feel the pinch but they will, ultimately, cease to produce that which will not *sell.*

We *can* control our population and thus decrease our per capita use of power. Population may be stabilized, and use of power reduced to what is necessary for a high quality of life. But population control will take time. We can begin now by ceasing to use power for trivial purposes.

Power use is presently divided about as follows in the United States: household and commercial, 33 percent; industrial, 42 percent; transportation, 24 percent. We must decide which uses, within each category, improve the quality of peoples' lives sufficiently to justify the inevitable pollution that results from power generation and use.

Household and Commercial

The term "standard of living" as used by utility spokesmen in the United States today generally means abundant luxuries, such as the following, for the affluent: electric blenders, toothbrushes and can openers, power saws, toys and mowers, dune buggies, luxury cars and golf carts, electric clothes dryers and garbage grinders, air conditioners, electric blankets and hair dryers.

Are these necessary for a high quality of life? We must realize that a decision made to purchase one of these "conveniences" is also a decision to accept the environmental deterioration that results from the production, use and disposal of the "convenience." Hand-operated blenders, toothbrushes, can openers and saws, clotheslines, blankets, bicycles, and feet produce much less pollution than the powered equivalents.

We can make the ecologically sensible decision to reject the concept of increasing perpetually the "standard of living" regardless of the human or ecological consequences. We can replace the outmoded industrial imperative—the "standard of living" concept—by the more human "quality of life" concept.

Many of us feel that the quality of our lives would be higher with far less use of energy in this country. We would be happy to do with fewer cars, substituting a transportation system that can make us mobile without dependence on the expensive, polluting, and dangerous automobile. We would be happy to see the last of glaring searchlights, neon signs, noisy power mowers and private airplanes, infernally noisy garbage trucks, dune buggies, and motorcycles. The quality of our lives is improved by each power plant not constructed near our homes or recreation areas, by each dam not constructed on a river used for canoeing. Quality of life is a positive ethic. Peace and quiet and fresh air are positive values; noisy smoking machines are negative ones.

Industry

Industry has been rapidly increasing its use of energy to increase production. An *Electrical World* pamphlet cheerfully describes this trend as follows:

> Industry's use of electric power has been increasing rapidly, too. The index of consumer use of electricity is kilowatts-per-hour. Industry's use is measured as the amount used per employee. Ten years ago, American industry used 24,810 kilowatt hours per year for each person employed. Today, the figure is estimated at 37,912. As industry finds more ways to use power to improve production, the output and wages of the individual employee rise.

Since unemployment is a problem and power use causes pollution, perhaps automation which uses energy to replace workers isn't a very good idea. Of course we could have full employment, a shorter work week, and less power use if we just wouldn't bother producing things that don't really improve the quality of life.

Transportation

If you wanted to design a transportation system to waste the earth's energy reserves and pollute the air as much as possible, you couldn't do much better than our present system dominated by the automobile. Only by following the advice of the popular science journals, placing in every garage a helicopter (using three times as much gasoline per passenger mile as a car) could you manage to do greater environmental damage.

Compared to a bus, the automobile uses from four to five times as much fuel per passenger mile. Compared to a train, it uses ten times as much. Walking and bicycling, of course, require no fuel at all.

Switching from the system of automobilism to a system of rapid transit, with more bicycling and walking in cities, would reduce fossil fuel consumption for transportation by a factor of almost 10. As transportation now accounts for 24 percent of the fuel expended, a saving of even 50 percent in this category would be helpful in reducing the rapid consumption of fossil fuels. Added benefits would be fewer deaths and injuries by automobiles, which have much higher injury rates than any form of public transportation; the liberation of much of the cities' space presently dedicated to the automobile; and less smog.

The term "standard of living" usually seems to apply only to Americans, and usually just to the present generation. It is important to think of all people in the world, and of future generations. The question must be asked whether it is fair to the rest of the world for the United States to use up such a disproportionate share of the world's energy resources. Even looking solely to United States interests, is it the best policy to use up our allotment as fast as possible?

If the whole world had equal rights and everyone burned fuel as fast as the U. S., the reserves would be gone very soon. The U. S. per capita rate of use of fossil fuels is from ten to a hundred times as great as the majority of people (the Silent Majority?) who live in the underdeveloped countries.

Each person in India uses only 1/83 as much power as an American. India now has 500,000,000 people or 2½ times the population of the U. S. Yet since each person uses so much less power, India's total power use is only 1/33 of that of the U.S. Its fair share would be 2½ times as much power as the U. S. The same argument, with somewhat different figures, holds for China, Southeast Asia, Pakistan, the Middle East, South America and Africa.

Not only does the burning of fossil fuels produce local pollution, but it also increases the carbon dioxide-to-oxygen ratio in the atmosphere. This occurs because each molecule of oxygen consumed in burning fuels results in the production of a carbon dioxide molecule (CH_2O plus O_2 yields CO_2 plus H_2O). This has the doubly adverse effect of taking oxygen out of the atmosphere, and putting carbon dioxide in, in equal amounts. The latter effect is of most concern to us because the CO_2 percentage in the atmosphere is minute compared to the huge reservoir of oxygen. While the atmosphere contains 20 percent oxygen, it has only 0.02 percent CO_2. Thus, fuel combustion reducing the O_2 concentration by only 1 percent would simultaneously increase the CO_2 concentration *tenfold.*

Each year the burning of fossil fuels produces an amount of carbon dioxide equal to about 0.5 percent of the existing carbon dioxide reservoir in the atmosphere. Of this production, half stays in the atmosphere, resulting in a 0.25 percent increase in atmospheric CO_2 per year. Of the other half, some becomes bound up with calcium or magnesium to become limestone, some becomes dissolved in the sea, and some is stored as the bodies of plants that fall to the deep, oxygen-poor sediments of the ocean and do not decompose.

If no CO_2 were being disposed of by the physical and biological processes in the ocean, then the CO_2 concentration of the atmosphere would increase by twice the present rate, because all of the CO_2 produced each year would remain in the atmosphere.

Burning all the recoverable reserves of fossil fuels would produce three times as much carbon dioxide as is now present in the atmosphere. If the present rate of increase in fuel use continues, and the rate of CO_2 dispersal continues unchanged, there will be an increase of about 170 percent in the CO_2 level in the next 150 years (which is the minimum estimate of the amount of time our fossil fuels will last). If the fuels last longer, say up to the "optimistic" 400 years that some predict, we will have that much more CO_2 increase, with the attendant smog and oil spills.

Scientists are becoming worried about increasing CO_2 levels because of the greenhouse effect, with its possible repercussions on the world climate. Most of the sun's energy striking the earth's surface is in the form of visible and ultraviolet rays from the sun. Energy leaves the earth as heat radiation or infrared rays. Carbon dioxide absorbs infrared rays more strongly than visible or ultraviolet rays. Energy coming toward the earth's surface thus readily passes through atmospheric carbon dioxide, but some escaping heat energy is absorbed and trapped in the atmosphere by carbon dioxide, much as heat is trapped in a greenhouse. This effect of carbon dioxide on the earth's climate has, in fact, been called the "greenhouse effect." Scientists differ in their opinions as to the eventual result this will have on our climate. Some believe that the earth's average temperature will increase, resulting in the melting of polar ice caps with an accompanying increase of sea levels and inundation of coastal cities. Others feel that there will be a temporary warming and partial melting of polar ice, but then greater evaporation from the open Arctic seas will cause a vast increase in snowfall, with an ensuing ice age.

Many people believe that green plants can produce a surplus of oxygen to compensate for that converted to CO_2 in burning fuels. This is not true. A plant produces only enough oxygen for its own use during its life plus enough extra for the oxidation of the plant after death to its original buildings blocks (CO_2 plus H_2O). Whether this oxidation occurs by fire, by bacterial decay, or by respiration of an animal eating the plant, has no effect on the ultimate outcome. When the plant is totally consumed by any of these three means, all of the oxygen it produced over its life is also consumed. The only way a plant leaves an oxygen surplus is if it fails to decompose, a relatively rare occurrence.

The important point is that fossil fuel combustion results in a change in the ratio of carbon dioxide to oxygen in the atmosphere, whereas use of oxygen by animals does not. This point is not generally understood, so two examples are discussed below.

First, since 70 percent of the world's oxygen is produced in the ocean, it has been forecast that death of the plankton in the ocean would cause asphyxiation of the animals of the earth. This is not the case because oxygen and carbon dioxide cycle in what is called the carbon cycle. A plant, be it a redwood tree or an algal cell, produces just enough oxygen to be used in consuming its carcass after death. The ocean plankton now produce 70 percent of the oxygen, but animals in the ocean use it up in the process of eating the plants. Very little of it is left over. The small amount that is left over is produced by plankton that have dropped to the oxygen-poor deep sediments and are essentially forming new fossil fuel.

If the plankton in the ocean were all to die tomorrow, all of the animals in the ocean would starve. The effect of this on the world's oxygen supply would be very small. The effect on the world's food supply, however, would be catastrophic. A large number of nations rely significantly on the ocean for food, particularly for high-quality protein. Japan, for example, is very heavily dependent on fisheries to feed itself.

Second, fears about reducing the world's oxygen supply have been expressed in reference to the cutting down of large forest areas, particularly in the tropics, where the soil will become hardened into bricklike laterite and no plant growth of any sort will be possible in the future. It will be a disaster if the Amazon rain forest is turned into laterite because the animals and people dependent on it could not exist. But this would have no effect on the world's oxygen balance. If the Amazon Basin were simply bricklike laterite, the area would produce no oxygen and consume no oxygen. At present the Amazon Basin is not producing surplus organic material. The same amount of organic material is present in the form of animal bodies, trees, stumps, and humus from year to year; therefore no net production of oxygen exists. The oxygen produced in the forest each year, which obviously is a large amount, is used up by the animals and microorganisms living in the forest in the consumption of the plant material produced over the preceding year.

In summary, I suggest that one goal of the environmental movement should be the reduction of total energy use in this country by 25 percent over the next decade. By doing this, we will have made a start toward preventing possibly disastrous climatic changes due to CO_2 buildup and the greenhouse effect. We will so reduce the need for oil that we can leave Alaska as wilderness and its oil in the ground. We will be able to stop offshore drilling with its ever-present probability of oil slick disasters, and won't need new supertankers which can spill more oil than the *Torrey Canyon* dumped on the beaches of Britain and France. We will be able to do without the risks of disease and accident from nuclear power plants. We won't need to dam more rivers for power. And perhaps most important, we can liberate the people from the automobile, whose exhausts turn the air over our cities oily brown (which causes 50,000 deaths a year) and which is turning our landscape into a sea of concrete. . . .

Robert Heilbroner: Ecological Armageddon

Ecology has become the Thing. There are ecological politics, ecological jokes, ecological bookstores, advertisements, seminars, teach-ins, buttons. The automobile, symbol of ecological abuse, has been tried, sentenced to death, and formally executed in at least two universities (complete with burial of one victim). Publishing companies are fattening on books on the sonic boom, poisons in the things we eat, perils loose in the garden, the dangers of breathing. The *Saturday Review* has appended a regular monthly Ecological Supplement. In short, the ecological issue has assumed the dimensions of a vast popular fad, for which one can predict with reasonable assurance the trajectory of all such fads—a period of intense popular involvement, followed by growing boredom and gradual extinction, save for a diehard remnant of the faithful.

This would be a great tragedy, for I have slowly become convinced during the last twelve months that the ecological issue is not only of primary and lasting importance, but that it may indeed constitute the most dangerous and difficult challenge that humanity has ever faced. Since these are very large statements, let me attempt to substantiate them by drawing freely on the best single descriptive and analytic treatment of the subject that I have yet seen, *Population, Resources, Environment,* by Paul and Anne Ehrlich of Stanford University. Rather than resort to the bothersome procedure of endlessly citing their arguments in quotation marks, I shall take the liberty of reproducing their case in a rather free paraphrase, as if it were my own, until we reach the end of the basic argument, after which I shall make clear some conclusions that I believe lie implicit, although not quite overt, in their work.

Ultimately, the ecological crisis represents our belated awakening to the fact that we live on what Kenneth Boulding has called, in the perfect phrase, our Spaceship Earth. As in all spaceships, sustained life requires that a meticulous balance be maintained between the life-support capability of the vehicle and the demands made by the inhabitants of the craft. Until quite recently, those demands have been well within the capability of the ship, both in terms of its ability to supply the physical and chemical requirements for continued existence and to absorb the waste products of the voyagers. This is not to say that the earth has been generous—short rations have been the lot of mankind for most of its history—nor is it to deny the recurrent advent of local ecological crises: witness the destruction of whole areas like the erstwhile granaries of North Africa. But famines have passed and there have always been new areas to move to. The idea that the earth as a whole was overtaxed is one that is new to our time.

For it is only in our time that we are reaching the ceiling of earthly carrying capacity, not on a local but on a global basis. Indeed, as will soon become clear, we are well past that capacity, provided that the level of resource intake and waste output represented by the average American or European is taken as a standard to be achieved by all humanity. To put it bluntly, if we take as the price of a first-class ticket the resource requirements of those passengers who travel in the Northern Hemisphere of the Spaceship, we have now reached a point at which the steerage is condemned to live forever—or at least within the horizon of the technology presently visible—at a second-class level; or at which a considerable change in living habits must be imposed on first class if the ship is ever to be converted to a one-class cruise.

This strain on the carrying capacity of the vessel results from the contemporary confluence of three distinct developments, each of which places tremendous or even unmanageable strains on the life-carrying capability of the planet and all of which together simply overload it. The first of these is the enormous strain imposed by the sheer burgeoning of population. The statistics of population growth are by now very well known: the earth's passenger list is growing at a rate that will give us some four billion humans by 1975, and that threatens to give us eight billions by 2010. I say "threatens," since it is likely that the inability of the earth to carry so large a group will result in an actual population somewhat smaller than this, especially in the steerage, where the growth is most rapid and the available resources least plentiful.

We shall return to the population problem later. But meanwhile a second strain is placed on the earth by the simple cumulative effect of *existing* technology (combustion engines, the main industrial processes, present-day agricultural techniques, etc.). This strain is localized mainly in the first-class portions of the vessel where each new arrival on board is rapidly given a standard complement of capital equipment and where the rate of physical- and chemical-resource transformation per capita steadily mounts. The strain consists of the limited ability of the soil, the water, and the atmosphere of these favored regions to absorb the outpourings of these fast-growing industrial processes.

The most dramatic instance of this limited absorptive power is the rise in the carbon dioxide content of the air due to the steady growth of (largely industrial) combustion. By the year 2000, it seems beyond dispute that the CO_2 content of the air will have doubled, raising the heat-trapping properties of the atmosphere. This so-called greenhouse effect has been predicted to raise main global temperatures sufficiently to bring catastrophic potential consequences. One possibility is a sequence of climatic changes resulting from a melting of the Arctic ice floes that would result in the advent of a new Ice Age; another is the slumping of the Antarctic icecap into the sea with a consequent tidal wave that could wipe out a substantial portion of mankind and raise the sea level by sixty to a hundred feet.

These are all "iffy" scenarios whose present significance may be limited to alerting us to the immensity of the ecological problem; happily they are of sufficient uncertainty not to cause us immediate worry (it is lucky they are, because it is extremely unlikely that all the massed technological and human energy on earth could arrest such changes once they began). Much closer to home is the burden placed on the earth's carrying capacity by the sheer requirements of a spreading industrial activity in terms of the fuel and mineral resources needed to maintain the going rate of output per person in the first-class cabins. To raise the existing (not the anticipated) population of the earth to American standards would require the annual extraction of great multiples of the quantities of iron, copper, lead, tin, etc., that we now take from the earth. Only the known reserves of iron allow us seriously to entertain the possibility of long-term mineral extraction at the required rates (and the capital investment needed to bring about such mining operations is enormous). And, to repeat, we have taken into account only today's level of population: to equip the prospective passengers of the year 2010 with this amount of basic raw materials would require a doubling of all the above figures.

I will revert later to the consequences of this prospect. First, however, let us pay attention to the third source of overload, this one traceable to the special environment-destroying potential of newly developed technologies. Of these the most important—and if it should ever come to full-scale war, of course the most lethal—is the threat posed by nuclear radiation. I shall not elaborate on this well-known (although not well-believed) danger, pausing to point out only that a massive nuclear holocaust would in all likelihood exert its principal effect in the Northern Hemisphere. The survivors in the South would be severely hampered in their efforts at reconstruction not only because most of the easily available resources of the world have already been used up, but because most of the technological know-how would have perished along with the populations up North.

But the threats of new technology are by no means limited to the specter of nuclear devastation. There is, immediately at hand, the known devastation of the new chemical pesticides that have now entered more or less irreversibly into the living tissue of the world's population. Most mothers' milk in the United States today—I now quote the Ehrlichs verbatim—"contains so much DDT that it would be declared illegal in interstate commerce if it were sold as cow's milk"; and the DDT intake of infants around the world is twice the daily allowable maximum set by the World Health Organization. We are already, in other words, being exposed to heavy dosages of chemicals whose effects we know to be dangerous, with what ultimate results we shall have to wait nervously to discover. (There is food for thought in the archeological evidence that one factor in the decline of Rome was the systematic poisoning of upper-class Romans from the lead with which they lined their wine containers.)

But the threat is not limited to pesticides. Barry Commoner predicts an agricultural crisis in the United States within fifty years from the action of our fertilizers, which will either ultimately destroy soil fertility or lead to pollution of the national water supply. At another corner of the new technology, the SST threatens not alone to shake us with its boom but to affect the amount of cloud cover (and climate) by its contrails. And I have not even mentioned the standard pollution problems of smoke, industrial effluents into lakes and rivers, or solid wastes. Suffice it to report that a 1968 UNESCO conference concluded that man has only about twenty years to go before the planet starts to become uninhabitable because of air pollution alone. Of course, "starts to" is imprecise; I am reminded of a cartoon of an industrialist looking at his billowing smokestacks, in front of which a forlorn figure is holding up a placard that says, "We have only 35 years to go." The caption reads, "Boy, that shook me up for a minute. I thought it said 3–5 years."

I have left until last the grimmest and gravest threat of all, speaking now on behalf of the steerage. This is the looming inability of the great green earth to bring forth sufficient food to maintain life, even at the miserable threshold of subsistence at which it is now endured by perhaps a third of the world's population. The problem here is the very strong likelihood that population growth will inexorably outpace whatever improvements in fertility and productivity we will be able to apply to the earth's mantle (including the watery fringes of the ocean where sea "farming" is at least technically imaginable). Here the race is basically between two forces: on the one hand, those that give promise that the rate of increase of population can be curbed (if not totally halted); and on the other, those that give promise of increasing the amount of sustenance we can wring from the soil.

Both these forces are subtly blended of technological and social factors. Take population growth. The great hope of every ecologist is that an effective birth-control technique—cheap, requiring little or no medical supervision, devoid of taboos or religious hindrances—will rapidly and effectively lower the present fertility rates which are doubling world population every 35 years (every 28 years in Africa; every 24 in Latin America). No such device is currently available, although the Pill, the IUD, vasectomies, abortions, condoms, coitus interruptus, and other known techniques could, of course, do the job if the requisite equipment, persuasion (or coercion), instruction, etc., could be brought to the 80 to 90 percent of the world's people who know next to nothing about birth control.

It seems a fair conclusion that no such worldwide campaign is apt to be successful for at least a decade and maybe a generation, although there is always the hope that a "spontaneous" change in attitudes similar to that in Hungary or Japan will bring about a rapid halt to population growth. But even in this unlikely event, the sheer "momentum" of population growth still poses terrible problems. Malcom Potts, Medical Director of International Planned Parenthood, has presented a shocking statistical calculation in this regard: he has pointed out that population growth in India is today adding a million mouths per month to the Indian subcontinent. If, by some miracle, fertility rates were to decline tomorrow by 50 percent in India, at the end of twenty years, owing to the already existing huge numbers of children who would be moving up into child-bearing ages, population growth in India would still be taking place at the rate of a million per month.

The other element in the race is our ability to match population growth with food supplies, at least for a generation or so, while birth-control techniques and campaigns are perfected. Here the problem is also partly technological, partly social. The technological part involves the so-called Green Revolution—the development of seeds that are capable, at their best, of improving yields per acre by a factor of 300 percent, sometimes even more. The problem, however, is that these new seeds generally require irrigation and fertilizer to bring their benefits. If India alone were to apply fertilizer at the per capita level of the Netherlands, she would consume half the world's total output of fertilizer. This would require a hundredfold expansion of India's present level of fertilizer use. Irrigation, the other necessary input for most improved seeds, poses equally formidable requirements. E. A. Mason of the Oak Ridge National Laboratories has prepared preliminary estimates of the costs of nuclear-powered "agro-industrial complexes" in which desalted water and fertilizer would be produced for use on adjacent farms. It would require 23 such plants per year, each taking care of some three million people, just to keep pace with present population growth. Since it would take at least five years to get these plants into operation, we should begin work today on at least 125 such units. Assuming that no hitches were encountered and that the technology on paper could be easily translated into a technology *in situ,* the cost would amount to $315 billion.

There are, as well, other technical problems associated with the Green Revolution of an ecological nature—mainly the risk of introducing locally untried strains of plants that may be subject to epidemic disease. But putting those difficulties to the side, we must recognize as well the social obstacles that a successful Green Revolution must overcome. The new seeds can only be afforded by the upper level of peasantry—not merely because of their cost (and the cost of the required fertilizer) but because only a rich peasant can take the risk of having the crop turn out badly without himself suffering starvation. Hence the Green Revolution is likely to increase the strains of social stratification within the underdeveloped areas. Then, too, even a successful local crop does not always shed its benefits evenly across a nation but results all too often in local gluts that cannot be transported to starving areas because of transportation bottle-necks.

None of these discouraging remarks are intended in the slightest to disparage the Green Revolution, which represents the inspired work of dedicated men. But the difficulties must be kept in mind as a corrective to the lulling belief that "science" can easily offset the population boom with larger supplies of food. There is no doubt that supplies of food *can* be substantially increased—rats alone devour some 10 to 12 percent of India's crop, and insects can ravage up to half the stored crops of some underdeveloped areas, so that even very "simple" methods of improved storage hold out important prospects of improving basic life-support, quite aside from the longer-term hopes of agronomy. Yet, at best these improvements will only stave off the day of reckoning. Ultimately the problem posed by Malthus must be faced—that population tends to increase geometrically, by doubling, and that agriculture does not, so that eventually population *must* face the limit of a food barrier. It is worth repeating the words of Malthus himself in this regard:

> Famine seems to be the last, the most dreadful resource of nature. The power of population is so much superior to the power in the earth to produce subsistence for man, that premature death must in some shape or other visit the human race. The vices of mankind are active and able ministers of depopulation . . . [S]hould they fail in this war of extermi-

> nation, sickly seasons, epidemics, pestilence, and plague, advance in terrific array, and sweep off their thousands and ten thousands. Should success still be incomplete, gigantic inevitable famine stalks in the rear, and with one mighty blow, levels the population with the food of the world.

This Malthusian prophecy has been so often "refuted," as economists have pointed to the astonishing rates of growth of food output in the advanced nations, that there is a danger of dismissing the warnings of the Ehrlichs as merely another premature alarm. To do so would be a fearful mistake. For, unlike Malthus, who assumed that technology would remain constant, the Ehrlichs have made ample allowance for the growth of technological capability, and their approach to the impending catastrophe is not shrill. They merely point out that a mild version of the Malthusian solution is already upon us, for at least half a billion people are chronically hungry or outright starving, and another one and a half billion under- or mal-nourished. Thus we do not have to wait for "gigantic inevitable famine"; it has already come.

What is more important is that the Ehrlichs see the matter in a perspective fundamentally different from Malthus', not as a problem involving supply and demand but as one involving a total ecological equilibrium. The crisis, as the Ehrlichs see it, is thus both deeper and more complex than merely a shortage of food, although the latter is one of its more horrendous evidences. What threatens the Spaceship Earth is a profound imbalance between the totality of systems by which human life is maintained and the totality of demands, industrial as well as agricultural, technological as well as demographic, to which that life-support capacity is subjected.

I have no doubt that one can fault bits and pieces of the Ehrlichs' analysis, and there is a note of determined pessimism in their work that leads me to suspect (or at least hope) that there is somewhat more time for adaptation than they suggest. Yet I do not see how their basic conclusion can be denied. Beginning within our lifetimes and rising rapidly to crisis proportions in our children's, humankind faces a challenge comparable to none in its history, with the possible exception of the forced migrations of the Ice Age. It is with the responses to this crisis that I wish to end this essay, for, telling and courageous as the Ehrlichs' analysis is, I do not believe that even they have fully faced up to the implications that their own findings present.

The first of these I have already stated: it is the clear conclusion that the underdeveloped countries can *never* hope to achieve parity with the developed countries. Given our present and prospective technology, there are simply not enough resources to permit a "Western" rate of industrial exploitation to be expanded to a population of four billion—much less eight billion—persons. It may well be that most of the population in the underdeveloped world has no ambition to reach Western standards—indeed, does not even know that such a thing as "development" is on the agenda. But the elites of these nations, for all their rhetorical rejection of Western (and especially American) styles of life, do tend to picture a Western standard as the ultimate end of their activities. As it becomes clear that such an objective is impossible, a profound reorientation of views must take place within the underdeveloped nations.

What such a reorientation will be it is impossible to say. For the near-term future, the outlook for the most population-oppressed areas will be a continuous battle against food shortages, coupled with the possible impairment of the intelligence of much of the surviving population due to protein deficiencies in childhood. This pressure of population may lead to aggressive searches for *Lebensraum,* or, as I have frequently written, it may culminate in revolutions of desperation. In the long run, of course, there is the possibility of considerable growth (although nothing resembling the attainment of a Western standard of consumption). But no quick substantial improvement in their condition seems feasible within the next generation at least. The visions of Sir Charles Snow or Soviet Academician Sakharov for a gigantic transfer of wealth from the rich nations to the poor (20 percent of GNP is proposed) are simply fantasies. Since much of GNP is spatially nontransferable or inappropriate, such a massive levy against GNP would imply shipments of up to 50 percent of much movable output. How this enormous flood of goods would be transported, allocated, absorbed, or maintained—*not to mention relinquished by the donor countries*—is nowhere analyzed by the proponents of such massive aid.

The implications of the ecological crisis for the advanced nations are not any less severe, although they are of a different kind. For it is clear that free industrial growth is just as disastrous for the Western nations as free population growth for those of the East and South. The worship in the West of a growing Gross National Product must be recognized as not only a deceptive but a very dangerous avatar; Kenneth Boulding has begun a campaign, in which I shall join him, to label this statistical monster Gross National Cost.

The necessity to bring our economic activities into a sustainable relationship with the resource capabilities and waste-absorption properties of the world will pose two problems for the West. On the simpler level, a whole series of technological problems must be met. Fume-free transportation must be developed on land and in the air. The cult of disposability must be replaced by that of reusability. Population stability must be attained through tax and other inducements, both to conserve resources and to preserve reasonable population densities. Many of these problems will tax our ingenuity, technical and socio-political, but the main problem they pose is not whether, but *how soon,* they can be solved.

But there is another, deeper question that the developed nations face—at least those that have capitalist economies. This problem can be stated as a crucial test as to who was right—John Stuart Mill or Karl Marx. Mill maintained, in his *Principles of Economics,* that the terminus of capitalist evolution would be a stationary state, in which the return to capital had fallen to insignificance, and a redistributive tax system would be able to capture any flows of income to the holders of scarce resources, such as land. In effect, he prophesied the transformation of capitalism, in an environment of abundance, into a balanced economy, in which the capitalist, both as the generator of change and as the main claimant on the surplus generated by change, would in fact undergo a painless euthanasia.

The Marxian view is, of course, quite the opposite. The very essence of capitalism, according to Marx, is expansion—which is to say, the capitalist, as a historical "type," finds his raison d'être in the insatiable search for additional money-wealth gained through the constant growth of the economic system. The idea of a "stationary" capitalism is, in Marxian eyes, a contradiction in terms, on a logical par with a democratic aristocracy or an industrial feudalism.

Is the Millian or the Marxian view correct? I do not think that we can yet say. Some economic growth is certainly compatible with a stabilized rate of resource use and disposal, for growth could take the form of the expenditure of additional labor on the improvement (aesthetic or technical) of the national environment. Indeed, insofar as education or cultural activity are forms of national output that require little resource use and result in little waste product, national output could be indefinitely expanded through these and similar activities. But there is no doubt that the main avenue of traditional capitalist accumulation would have to be considerably constrained; that net investment in mining and manufacturing would likely decline; that the rate and kind of technological change would need to be supervised and probably greatly reduced; and that, as a consequence, the flow of profits would almost certainly fall.

Is this imaginable within a capitalist setting—that is, in a nation in which the business ideology permeates the views of nearly all groups and classes, and establishes the bounds of what is possible and natural, and what is not? Ordinarily I do not see how such a question could be answered in any way but negatively, for it is tantamount to asking a dominant class to acquiesce in the elimination of the very activities that sustain it. But this is an extraordinary challenge that may evoke an extraordinary response. Like the challenge posed by war, the ecological crisis affects all classes, and therefore may be sufficient to induce sociological changes that would be unthinkable in ordinary circumstances. The capitalist and managerial classes may see—perhaps even more clearly than the consuming masses—the nature and nearness of the ecological crisis, and may recognize that their only salvation (as human beings, let alone privileged human beings) is an occupational migration into governmental or other posts of power, or they may come to accept a smaller share of the national surplus simply because they recognize that there is no alternative. When the enemy is nature, in other words, rather than another social class, it is at least imaginable that adjustments could be made that would be impossible in ordinary circumstances.[1]

There is, however, one last possibility to which I must also call attention. It is the possibility that the ecological crisis will simply result in the decline, or even destruction, of Western civilization, and of the hegemony of the scientific-technological view that has achieved so much and cost us so dearly. Great challenges do not always bring great responses, especially when those responses must be sustained over long periods of time and require dramatic changes in life-styles and attitudes. Even educated men today are able to deny the reality of the crisis they face: there is wild talk of farming the seas, of transporting men to the planets, of unspecified "miracles" of technology that will avert disaster. Glib as they are, however, at least these suggestions have a certain responsibility when compared with another and much more worrisome response: *je m'en fiche*. Can we really persuade the citizens of the Western world who are just now entering the heady atmosphere of a high-consumption way of life that conservation, stability, frugality, and a deep concern for the distant future must take priority over the personal indulgence for which they have been culturally prepared and which they are about to experience for the first time? Not the least danger of the ecological crisis, as I see it, is that tens and hundreds of millions will shrug their shoulders at the prospects ahead ("What has posterity ever done for us?"), and that the increasingly visible approach of ecological Armageddon will bring not repentance but Saturnalia.

Yet I cannot end this essay on such a note. For it seems to me that the ecological enthusiasts may be right when they speak of the deteriorating environment as providing the *possibility* for a new political rallying ground. If a new New Deal, capable of engaging both the efforts and the beliefs of this nation, is the last great hope to which we cling in the face of what seems otherwise to be an inevitable gradual worsening and coarsening of our style of life, it is possible that a determined effort to arrest the ecological decay might prove to be its underlying theme. Such an issue, immediate in the experience of all, carries an appeal that might allow vast improvements to be worked in the American environment, both urban and industrial. I cannot estimate the likelihood of such a political awakening, dependent as these matters are on the dice of personality and the outcome of events at home and abroad. But however slim the possibility of bringing about such a change, it does at least make the ecological crisis, unquestionably the gravest long-run threat of our times, potentially the source of its greatest short-term promise.

---

[1] Let me add a warning that it is not only capitalists who must make an unprecedented ideological adjustment. Socialists must also come to terms with the abandonment of the goal of industrial superabundance on which their vision of a transformed society rests. The stationary equilibrium imposed by the constraints of ecology requires at the very least a reformulation of the kind of economic society toward which socialism sets its course.

Paul and Anne Ehrlich: Population, Resources, Environment

The Epidemiological Environment

Today the population of *Homo sapiens* is the largest in the history of the species, it has the highest average density, and it contains a record number of undernourished and malnourished people. The population is also unprecedentedly mobile. People are in continual motion around the globe, and they are able to move from continent to continent in hours. The potential for a worldwide epidemic (pandemic) has never been greater, but people's awareness of this threat has probably never been smaller.

We do not completely understand the behavior of viruses but we do know that the spontaneous development of highly lethal strains of human viruses and the invasion of humanity by extremely dangerous animal viruses are possible. We also know that crowding increases the chances for development of a virus epidemic. Should, say, an especially virulent strain of flu appear, it is doubtful that the United States and other developed countries could produce enough vaccine fast enough to save most of their populations. Needless to say the problem would be even more severe in the UDCs. Certainly little effort could be made to save most of humanity. Consider, for example, the difficulty the United States had in coping with the mild Asian flu epidemic of 1968. It was not possible to manufacture enough vaccine to protect most of the population, and the influenza death rate in 1968 was more than four times as high as that of 1967. Only 613 deaths were attributed to flu, but society paid a high price for the disease in extra medical care and loss of working hours.

In 1967 an outbreak of a previously unknown disease occurred among a shipment of vervet monkeys that had been imported into laboratories in Marburg, Germany, and in Yugoslavia. This severe, hemorrhagic disease infected twenty-five laboratory workers who came into contact with the monkeys and their tissues. Seven of these people died. Five secondary infections occurred in individuals who came into contact with the blood of the original patients; all of these individuals survived. Mankind was

---

extremely fortunate that the first infections of *Homo sapiens* by Marburg-virus occurred around laboratories where the nature of the threat was quickly recognized, and the disease contained (it was not susceptible to antibiotics). If it had escaped into the human population at large, and if the disease had retained its virulence as it passed from person to person, an epidemic resulting in hundreds of millions or even billions of deaths might have occurred. Among well-fed laboratory workers with expert medical care, 7 out of 30 patients died. Among hungry people with little or no medical care, mortality would be much higher. The infected monkeys passed through London airport in transit to the laboratories. If the virus had infected airport personnel, it could have spread over the entire world before anyone realized what was happening.

Our highly mechanized society is also extremely vulnerable to disruption by such events as power failures, floods, and snowstorms. What would happen if the United States were confronted with an epidemic that kept masses of sick people from work and caused the uninfected to stay home or flee the cities because of their fear of infection? This might slow or even stop the spread of the disease, but hunger, cold (in the winter), and many other problems would soon develop as the services of society ceased to operate. We have substantial knowledge of the almost total breakdown of much less complex societies than ours in the face of the "Black Plague"—a breakdown that occurred among people far more accustomed to a short life, hardship, disease, and death than the population of the Western World today. The panic may well be imagined if Americans were to discover that "modern medical science" either had no cure for a disease of epidemic proportions, or had insufficient doses of the cure for everyone. The disease itself would almost certainly impede the application of any ameliorating measures. Distribution of vaccines, for instance, would be difficult if airlines, trains, and trucks were not running. . . .

Economic and Political Change

In relation to the population-resources-environment crisis, economics and politics can usually be viewed as two sides of the same coin. A very large number of political decisions are made on an economic basis, especially those relating to environmental problems. John Maynard Keynes wrote in *The General Theory of Employment, Interest, and Money* (1936): "The ideas of economists and political philosophers, both when they are right and when they are wrong, are more powerful than is commonly understood. Indeed the world is ruled by little else." Although the major political division of our time—that between capitalist and communist worlds—is thought to be based on differences in economic ideology, the actual differences are relatively few. In fact, a major cause of humanity's current plight lies not in the economic differences between the two superpowers, but in the economic attitudes that they have in common.

Gross National Product and Economic Growthmanship

Economists are not unanimous in their view of economic growth. Paul A. Samuelson wrote in *Economics, An Introductory Analysis* (1967): "The ghost of Carlyle should be relieved to know that economics, after all, has not been a dismal science. It has been the cheerful, but impatient, science of growth." On the other hand, E. J. Mishan states in *The Costs of Economic Growth* (1967):

> The skilled economist, immersed for the greater part of the day in pages of formulae and statistics, does occasionally glance at the world about him and, if perceptive, does occasionally feel a twinge of doubt about the relevance of his contribution. . . . For a moment, perhaps, he will dare wonder whether it is really worth it. Like the rest of us, however, the economist must keep moving, and since such misgivings about the overall value of economic growth cannot be formalized or numerically expressed, they are not permitted seriously to modify his practical recommendations.

The majority of economic theorists hold Samuelson's view and still tend to be growth-oriented, as do most politicians and businessmen in both DCs and UDCs.

In much of the world—indeed, in all countries with any aspirations towards "modernization," "progress," or "development"—a general economic index of advancement is growth of the gross national product (GNP). The GNP is the sum of personal and government expenditure on goods and services, plus expenditure on investment. More important than what the GNP is, however, is what it *is not.* It is not a measure of the degree of freedom of the people of a nation. It is not a measure of the health of a population. It is not a measure of the state of depletion of natural resources. It is not a measure of the stability of the environmental systems upon which life depends. It is not a measure of security from the threat of war. It is not, in sum, a comprehensive measure of the *quality* of life.

When the standard of living of two nations is compared, it is customary to examine their *per capita* GNPs. Per capita GNP is an especially unfortunate statistic. First of all it is the ratio of two statistics that are at best crude estimates, especially in the UDCs, where neither GNP nor population size is known with any accuracy. More important, comparisons of per capita GNP overestimate many kinds of differences. For instance, a comparison of per capita GNPs would lead to the conclusion that the average American lives almost ten times as well as the average Portuguese, and some sixty times as well as the average Burmese. This of course is meaningless, since virtually all services and many goods are much cheaper in the UDCs. Americans pay perhaps five or ten times as much for farm labor, domestic help, haircuts, carpentry, plumbing, and so forth as do people in the UDCs, and the services we get are often of inferior quality. And yet these services, because of the accounting system, contribute between five and ten times as much to our GNP as the same services do to the GNPs of, say, Burma or India. Furthermore, figures on the increase of per capita GNP in UDCs do not take into account such things as rise in literacy rate, and thus may underrate the amount of progress a country has made toward modernization.

Nor does the GNP measure many negative aspects of the standard of living. Although the average Burmese may live only one-sixtieth as well as the average American, the average American may cause a hundred times as much ecological destruction to the planet as a whole.

A serious criticism that can be leveled at the majority of economists is one that applies equally to most people and societies: they accept a doctrine of economic determinism. The myths of cornucopian economics as opposed to the realities of geology and biology have already been discussed, but the problem is much more pervasive than that. Economic growth has become *the* standard for progress, *the* benefit for which almost any social cost is to be paid. This problem in economic thought can be fully appreciated by a perusal of Samuelson's excellent *Economics,* one of the best and most influential textbooks ever written. The book is, of course, oriented towards economic growth. Problems of the growing scarcity of nonrenewable resources are presented only briefly as a problem of underdeveloped countries. The physical limits placed on economic growth by the thermal problems associated with energy consumption are not discussed in the text, nor are the other basic environmental constraints discussed in our earlier chapters. Implicit in the treatment of economic development is the idea that it is possible for 5 to 7 billion people to achieve a standard of living similar to that of the average American of the 1960's. Uninformed technological optimism is explicit or implicit throughout the book.

Nevertheless, Samuelson's book reveals more understanding of population and environment than the writings of many other economists. He does realize that growth of GNP must be "qualified by data on leisure, population size, relative distribution, quality, and noneconomic factors." Furthermore, in a recent *Newsweek* column (October 6, 1969) Samuelson wrote, ". . . most of us are poorer than we realize. Hidden costs are accruing all the time; and because we tend to ignore them, we overstate our incomes . . . Thomas Hobbes said that in the state of nature the life of man was nasty, brutish and short. In the state of modern civilization it has become nasty, brutish and long."

The discussion of the problems of UDCs in Samuelson's text is a model of realism when contrasted with the ideas of such economic conservatives as Milton Friedman of the University of Chicago. Most economists subscribe to the "bigger and more is better" philosophy. The growing mixed economy is something to analyze, improve, and by all means to keep growing. In an article that appeared in the *New York Review of Books,* economist Wassily Leontief of Harvard remarked that ". . . If the 'external costs' of growth clearly seem to pose dangers to the quality of life, there is as yet no discernible tendency among economists or economic managers to divert their attention from this single-minded pursuit of economic growth." That economists have clung to this idea is not surprising. After all, natural scientists often cling to outmoded ideas that have produced far less palpable benefits than have the mixed economies of the Western World. The question of whether a different economic system might have produced a more equitable *distribution* of benefits is not one that Western economists like to dwell on. Furthermore, ideas of perpetual growth are congruent with the conventional wisdom of most of the businessmen of the world; indeed, of most of the world's population. The people of the UDCs naturally wish to emulate the economic growth of the West and they long for "development" with all of its shiny accoutrements. Why should they be expected to know that it is physically and ecologically impossible for them to catch up with us when many of our most erudite citizens are still unaware of that fact?

Perhaps most serious is the common idea that not only is growth of the GNP highly desirable, but that population increase, at least in DCs, *promotes* such growth. However, there are some economists, such as J. J. Spengler, of the University of North Carolina, who have made a point of attacking the idea that population growth is necessary to keep the GNP growing in DCs. Certainly in the DCs there is no perfect correlation between population growth and growth of GNP. As Spengler says, "It is high time . . . that business cease looking upon the stork as a bird of good omen." He points out that a substantial portion of the GNP consists of services, and that these may continue to expand with a static population. Even in such heavily people-dependent industries as transportation, there is considerable room for attracting a greater proportion of the population to the use of the service. For instance, a small percentage of the American population does the major portion of airline traveling, and presumably airlines could grow for several generations even if the population size remained stable. . . .

But, whether or not population growth helps to raise the GNP, it is clear that the GNP cannot grow forever. Why should it? As John Kenneth Galbraith points out in *The New Industrial State,* it would be entirely logical to set limits on the amount of product a nation needs, and then to strive to reduce the amount of work required to produce such a product (and, one might add, to see that the product is much more equitably distributed than it is today). But, of course, such a program would be a threat to some of the most dearly held beliefs of our society. It would attack the Protestant ethic, which insists that one must be kept busy on the job for 40 hours a week. It is even better to work several more hours moonlighting, so that the money can be earned to buy all those wonderful automobiles, detergents, appliances, and assorted gimcracks which *must* be bought if the economy is to continue to grow. But this tradition is outmoded; the only hope for civilization in the future is to work for *quality* in the context of a nongrowing economy.

Economist Kenneth C. Boulding has begun to develop an exciting set of economic concepts dealing with the population-resource-environment crisis, by recognizing the existence of biological and physical limitations to growth. In "The Economics of the Coming Spaceship Earth," he described the need to shift from our present "cowboy economy," in which both production and consumption are regarded with great favor, and which is "associated with reckless, exploitative, romantic and violent behavior," to a "spaceman-economy." In the spaceman economy there are no unlimited reservoirs, either for extraction or pollution, and consumption must be minimized. In a classical understatement, Boulding describes the idea that production and consumption are bad things as "very strange to economists." But even economists can change, and perhaps this section on economics can end on an optimistic note. Economists of the next generation may be weaned away from their concentration on perpetual growth and high production-consumption and learn, in Boulding's words, to measure economic success in terms of the "nature, extent, quality and complexity of the total capital stock, including in this the state of the human bodies and minds included in the system."

Personal Freedom and the Quality of Life

> We have had no environmental index, no census statistics to measure whether the country is more or less habitable from year to year. A tranquility index or a cleanliness index might have told us something about the condition of man, but a fast-growing country preoccupied with making and acquiring material things has had no time for the amenities that are the very heart and substance of daily life.

Stewart Udall was Secretary of the Interior when he wrote those words. His bold challenge to Americans, expressed in *1976: Agenda for Tomorrow,* has not yet been accepted. Is there any way to break into the present system and persuade our society to weigh economic goals carefully against other possible goals of human existence? Can we proceed with Mr. Udall's urgent agenda? The obstacles are great, since economics and politics are so intertwined, and the various elements of the power structure in the U. S. all want and promote "growth." If there is any chance of getting a reversal of this attitude, it lies in convincing those in power as well as the electorate that their own personal lives and freedom are at stake. More men with dedication and perception must be elected to public office, and ways must be found to convince the present leaders of the nation that population growth and accelerating resource utilization, coupled with environmental decay, are injuring their children and progressively limiting their possible futures. Two points may be made:

1. While the American economy has been growing, freedom has been shrinking. Greater and greater controls must be applied to everyday living, and restrictions will inevitably grow more severe as population increases. The use of automobiles, boats, and private airplanes will become even more circumscribed. The keeping of pets, especially those that are noisy or consume substantial amounts of protein, will be forbidden. Increasingly, access to recreational areas will be strictly rationed. Bureaucracy will continue to grow as the government tries to solve more and more pressing problems with less and less success. Each citizen, assuming that at least an illusion of "democracy" will persist, will have an ever-decreasing say in the affairs of state. For instance, consider what has happened to the average citizen's "say" in the past 150 years. In 1810 each of 52 Senators represented, on the average, about 140,000 citizens. Today each of 100 Senators represents about 2,100,000 Americans—15 times as many! A parallel dilution of representation has occurred in the House of Representatives.
2. While the American GNP has been growing, the quality of life in the United States has been deteriorating. The GNP roughly doubled in the decade 1960–1969. Can anyone claim that the average individual's life has greatly improved in the same period? Here is a short list of the negative changes that have occurred in the quality of his existence. The air that he breathes has become more foul, and the quality of the water he uses has probably declined. His chances of being robbed or murdered have increased, as have his chances of losing his life in a highway accident or his home or business in a civil disorder. His chances of dying of emphysema, bronchitis, and various kinds of cancer have increased. He must travel further to reach solitude, either on increasingly crowded highways, increasingly shoddy trains, or increasingly delay-prone airlines. His children have a more difficult time being accepted into a first class college than they did in 1960. If he has sons they are more likely to be killed in a war, or to flee the country to avoid being drafted. Can a list of *improvements* twice this long easily be constructed?

We suspect that these arguments will not have the desired effect, even if a great many influential people get the message. To a large extent the political-economic system has a life of its own; it possesses emergent qualities beyond those of the individuals who constitute it. There is no conscious conspiracy on the part of individual military men, businessmen, and government officials to destroy the United States and the world, but the total effect of their actions and those of their counterparts in other governments nonetheless is moving us toward that end. People in groups, be they mobs, university committees, armies, or industrial boards, simply do not behave the way single individuals behave. But basically we must try, by changing the behavior of many individuals, to produce the desired changes in the corporate behavior of the economic and political establishment. . .

2 Arguments against Economic Growth

We have seen that a number of prominent ecologists consider continuous economic growth and development the most important cause of environmental destruction. Economists, on the other hand, have long accepted the social value of economic growth without question. This is not as true today, however, since a number of economists, as well as other social scientists, have begun to see less value in increasing abundance, especially when the social costs of obtaining it are considered.

Kenneth Boulding, a well-known economist, social scientist, and environmentalist, coined the already classic labels of "cowboy" and "spaceman" economies in his essay "The Economics of the Coming Spaceship Earth." He argues that it is not the flow of goods and services (GNP) that ought to be maximized, but rather the "nature, extent, quality, and complexity of the total capital stock, including in this the state of the human bodies and minds included in the system." It is the quality of life which is to be maximized, and this is not by any means attained by maximizing the Gross National Product. On the whole Boulding's point of view can be characterized as long-run in perspective and generally optimistic.

The contribution by Ezra Mishan is taken from his book *The Costs of Economic Growth,* probably the most explicit statement against economic growth that has been written by an economist of traditional outlook. Mishan is professor of welfare economics at the London School of Economics. He combats economic orthodoxy in terms of its own ground rules, in this case, welfare economics. He argues that advertising tends to decrease welfare since economic resources are spent primarily to create wants and dissatisfaction rather than to satisfy existing wants. It is interesting that as soon as economic growth is seen as undesirable, the major functional contribution of advertising, that of maintaining adequate levels of demand, is no longer relevant. Secondly, he suggests that in wealthy nations like the United States, it is only through increases in relative affluence—by getting farther ahead of the Joneses—that individual welfare can be increased. This "relative income hypothesis," to the extent it is true, undermines the whole basis for continued increases in abundance.

Since the subject of a no-growth economy is such an unorthodox area of study, there is a great range in the methods of analysis used as different individuals express their personal orientation, often working largely on their own. In "Toward a New Economics—Questioning Growth," Herman E. Daly of the Department of Economics at Louisiana State University starts from the classical economist's concept of the stationary state and explores some of the implications of this concept from a theoretical perspective. His study leads to conclusions about the inappropriateness of marginal analysis, of what would be necessary to encourage more leisure, the limits of recycling in an economy with a large flow of physical goods, and the loss of validity behind arguments in support of inequality of distribution in an economy where the maximization of production is no longer necessary. The question resolves ultimately into how the surplus can be distributed and consumed rather than reinvested, a problem that Daly does not take on.

The next paper is by three San Diego State College economics professors, John Hardesty, Norris Clement, and Clinton Jencks. "The Political Economy of Environmental Destruction" suggests that the surplus to be distributed may have to be significantly reduced, that it is the absolute size of the Gross National Product that is already causing the destruction of the environment. The authors argue that a reduction in GNP of at least fifty percent is necessary, as well as zero growth in the future, and then proceed to discuss some of the implications of such a drastic change in the functioning of the economy. They conclude that capitalism and zero GNP growth are incompatible and that major changes are called for in the direction of a noncapitalist, communal society, as well as slower paced, nonmaterialistic styles of life.

The other editor of this volume, Warren A. Johnson, tackles the question of distribution of income in "The Guaranteed Income as an Environmental Measure." The guaranteed income is seen as one way to escape the present need to maintain full employment, which requires economic growth to provide the necessary jobs. The author sees the guaranteed income as a device that has the potential to discourage the tendency to overproduction while encouraging the development of life styles that are not so dependent on economically productive work and high levels of consumption. The potential for this alternative seems enhanced by the limited proposals for income maintenance that are now being considered, such as President Nixon's Family Assistance proposal, as well as the difficulties that are already being encountered in providing jobs for the educated and uneducated alike. The author is a member of the Geography Department at San Diego State College with a degree in resource planning and conservation.

The next article, "Buddhist Economics," is offered as one alternative to our growth-oriented economics. The author, E. F. Schumacher, was born in Germany in 1911, went to England as a Rhodes Scholar, and emigrated there in 1937. He has been economic adviser to the governments of Burma and India, and his *Roots of Economic Growth* was published by the Ghandian Institute of Studies in India. In "Buddhist Economics" he suggests that there are other functions of work just as important as providing goods and services, functions that are often destroyed by specialization of labor, by economic integration, and by machines that take over the human part of the work. He argues that much greater human satisfaction can be found through the moderation of means that is characteristic of "The Middle Way" of Buddhism, with less violence to man and nature.

In the *Challenge of Man's Future,* Harrison Brown of the California Institute of Technology studies the future from the point of view of a geophysicist concerned with resources and the type of technology that will be necessary to provide for the anticipated material needs of the future. He includes in his analysis the type of society that will be needed to operate this "machine civilization." He makes two major points that cannot be overemphasized. First, as our industrial society continues to develop we must expect increasingly rigid organization, and possibly totalitarian control, to maintain the highly integrated and efficient economic processes necessary to obtain raw materials from such sources as sea water and granite. Second, if this "machine civilization" ever breaks down, say through nuclear war, it is unlikely to ever be regained, since essential resources are now attainable only by the most sophisticated means. It is not an attractive prospect, but it is very much the direction we take when we rely on technological solutions rather than social change. What our society has consistently failed to foresee is that technological change inevitably brings with it social change, and often in undesired forms.

The arguments presented against economic growth suggest that man is on an extremely hazardous path, at best toward something akin to Orwell's *1984* or Huxley's *Brave New World,* at worst toward some form of breakdown and chaos. Clearly, we must take action to leave this path. Taken in this context, the article "Congress and Pollution—The Gentleman's Agreement" is a depressing piece. The authors are Douglas Ross, legislative assistant to former Senator Joseph Tydings, and Harold Wolman, of the National Urban Coalition. If they are correct, and the political process cannot effectively handle the rising ecological crisis, then we must look for the means to change society outside the established political channels. This is no easy task, whether undertaken by evolutionary or revolutionary means.

Kenneth E. Boulding: The Economics of the Coming Spaceship Earth

We are now in the middle of a long process of transition in the nature of the image which man has of himself and his environment. Primitive men, and to a large extent also men of the early civilizations, imagined themselves to be living on a virtually illimitable plane. There was almost always somewhere beyond the known limits of human habitation, and over a very large part of the time that man has been on earth, there has been something like a frontier. That is, there was always some place else to go when things got too difficult, either by reason of the deterioration of the natural environment or a deterioration of the social structure in places where people happened to live. The image of the frontier is probably one of the oldest images of mankind, and it is not surprising that we find it hard to get rid of.

Gradually, however, man has been accustoming himself to the notion of the spherical earth and a closed sphere of human activity. A few unusual spirits among the ancient Greeks perceived that the earth was a sphere. It was only with the circumnavigations and the geographical explorations of the fifteenth and sixteenth centuries, however, that the fact that the earth was a sphere became at all widely known and accepted. Even in the nineteenth century, the commonest map was Mercator's projection, which visualizes the earth as an illimitable cylinder, essentially a plane wrapped around the globe, and it was not until the Second World War and the development of the air age that the global nature of the planet really entered the popular imagination. Even now we are very far from having made the moral, political, and psychological adjustments which are implied in this transition from the illimitable plane to the closed sphere.

Economists in particular, for the most part, have failed to come to grips with the ultimate consequences of the transition from the open to the closed earth. One hesitates to use the terms "open" and "closed" in this connection, as they have been used with so many different shades of meaning. Nevertheless, it is hard to find equivalents. The open system,

From *Environmental Quality in a Growing Economy,* ed. by Henry Jarrett, 1966. Published by The Johns Hopkins Press for Resources for the Future, Inc. Reprinted by permission of the publisher.

indeed, has some similarities to the open system of von Bertalanffy,[1] in that it implies that some kind of a structure is maintained in the midst of a throughput from inputs to outputs. In a closed system, the outputs of all parts of the system are linked to the inputs of other parts. There are no inputs from outside and no outputs to the outside; indeed, there is no outside at all. Closed systems, in fact, are very rare in human experience, in fact almost by definition unknowable, for if there are genuinely closed systems around us, we have no way of getting information into them or out of them; and hence if they are really closed, we would be quite unaware of their existence. We can only find out about a closed system if we participate in it. Some isolated primitive societies may have approximated to this, but even these had to take inputs from the environment and give outputs to it. All living organisms, including man himself, are open systems. They have to receive inputs in the shape of air, food, water, and give off outputs in the form of effluvia and excrement. Deprivation of input of air, even for a few minutes, is fatal. Deprivation of the ability to obtain any input or to dispose of any output is fatal in a relatively short time. All human societies have likewise been open systems. They receive inputs from the earth, the atmosphere, and the waters, and they give outputs into these reservoirs; they also produce inputs internally in the shape of babies and outputs in the shape of corpses. Given a capacity to draw upon inputs and to get rid of outputs, an open system of this kind can persist indefinitely.

There are some systems—such as the biological phenotype, for instance the human body—which cannot maintain themselves indefinitely by inputs and outputs because of the phenomenon of aging. This process is very little understood. It occurs, evidently, because there are some outputs which cannot be replaced by any known input. There is not the same necessity for aging in organizations and in societies, although an analogous phenomenon may take place. The structure and composition of an organization or society, however, can be maintained by inputs of fresh personnel from birth and education as the existing personnel ages and eventually dies. Here we have an interesting example of a system which seems to maintain itself by the self-generation of inputs, and in this sense is moving towards closure. The input of people (that is, babies) is also an output of people (that is, parents).

Systems may be open or closed in respect to a number of classes of inputs and outputs. Three important classes are matter, energy, and information. The present world economy is open in regard to all three. We can think of the world economy or "econosphere" as a subset of the "world set," which is the set of all objects of possible discourse in the world. We then think of the state of the econosphere at any one moment as being the

---

[1] Ludwig von Bertalanffy, *Problems of Life* (New York: John Wiley and Sons, 1952).

total capital stock, that is, the set of all objects, people, organizations, and so on, which are interesting from the point of view of the system of exchange. This total stock of capital is clearly an open system in the sense that it has inputs and outputs, inputs being production which adds to the capital stock, outputs being consumption which subtracts from it. From a material point of view, we see objects passing from the noneconomic into the economic set in the process of production, and we similarly see products passing out of the economic set as their value becomes zero. Thus we see the econosphere as a material process involving the discovery and mining of fossil fuels, ores, etc., and at the other end a process by which the effluents of the system are passed out into noneconomic reservoirs—for instance, the atmosphere and the oceans—which are not appropriated and do not enter into the exchange system.

From the point of view of the energy system, the econosphere involves inputs of available energy in the form, say, of water power, fossil fuels, or sunlight, which are necessary in order to create the material throughput and to move matter from the noneconomic set into the economic set or even out of it again; and energy itself is given off by the system in a less available form, mostly in the form of heat. These inputs of available energy must come either from the sun (the energy supplied by other stars being assumed to be negligible) or it may come from the earth itself, either through its internal heat or through its energy of rotation or other motions, which generate, for instance, the energy of the tides. Agriculture, a few solar machines, and water power use the current available energy income. In advanced societies this is supplemented very extensively by the use of fossil fuels, which represent as it were a capital stock of stored-up sunshine. Because of this capital stock of energy, we have been able to maintain an energy input into the system, particularly over the last two centuries, much larger than we would have been able to do with existing techniques if we had had to rely on the current input of available energy from the sun or the earth itself. This supplementary input, however, is by its very nature exhaustible.

The inputs and outputs of information are more subtle and harder to trace, but also represent an open system, related to, but not wholly dependent on, the transformations of matter and energy. By far the larger amount of information and knowledge is self-generated by the human society, though a certain amount of information comes into the sociosphere in the form of light from the universe outside. The information that comes from the universe has certainly affected man's image of himself and of his environment, as we can easily visualize if we suppose that we lived on a planet with a total cloud-cover that kept out all information from the exterior universe. It is only in very recent times, of course, that the information coming in from the universe has been captured and coded into the form of a complex image of what the universe is like outside the earth;

but even in primitive times, man's perception of the heavenly bodies has always profoundly affected his image of earth and of himself. It is the information generated within the planet, however, and particularly that generated by man himself, which forms by far the larger part of the information system. We can think of the stock of knowledge, or as Teilhard de Chardin called it, the "noosphere," and consider this as an open system, losing knowledge through aging and death and gaining it through birth and education and the ordinary experience of life.

From the human point of view, knowledge or information is by far the most important of the three systems. Matter only acquires significance and only enters the sociosphere or the econosphere insofar as it becomes an object of human knowledge. We can think of capital, indeed, as frozen knowledge or knowledge imposed on the material world in the form of improbable arrangements. A machine, for instance, originated in the mind of man, and both its construction and its use involve information processes imposed on the material world by man himself. The cumulation of knowledge, that is, the excess of its production over its consumption, is the key to human development of all kinds, especially to economic development. We can see this pre-eminence of knowledge very clearly in the experiences of countries where the material capital has been destroyed by a war, as in Japan and Germany. The knowledge of the people was not destroyed, and it did not take long, therefore, certainly not more than ten years, for most of the material capital to be reestablished again. In a country such as Indonesia, however, where the knowledge did not exist, the material capital did not come into being either. By "knowledge" here I mean, of course, the whole cognitive structure, which includes valuations and motivations as well as images of the factual world.

The concept of entropy, used in a somewhat loose sense, can be applied to all three of these open systems. In the case of material systems, we can distinguish between entropic processes, which take concentrated materials and diffuse them through the oceans or over the earth's surface or into the atmosphere, and anti-entropic processes, which take diffuse materials and concentrate them. Material entropy can be taken as a measure of the uniformity of the distribution of elements and, more uncertainly, compounds and other structures on the earth's surface. There is, fortunately, no law of increasing material entropy, as there is in the corresponding case of energy, as it is quite possible to concentrate diffused materials if energy inputs are allowed. Thus the processes for fixation of nitrogen from the air, processes for the extraction of magnesium or other elements from the sea, and processes for the desalinization of sea water are anti-entropic in the material sense, though the reduction of material entropy has to be paid for by inputs of energy and also inputs of information, or at least a stock of information in the system. In regard to matter, therefore, a closed system is conceivable, that is, a system in which there is neither increase nor decrease in material entropy. In such a system all outputs from consumption would constantly be recycled to become inputs for production, as for instance, nitrogen in the nitrogen cycle of the natural ecosystem.

In regard to the energy system there is, unfortunately, no escape from the grim Second Law of Thermodynamics; and if there were no energy inputs into the earth, any evolutionary or developmental process would be impossible. The large energy inputs which we have obtained from fossil fuels are strictly temporary. Even the most optimistic predictions would expect the easily available supply of fossil fuels to be exhausted in a mere matter of centuries at present rates of use. If the rest of the world were to rise to American standards of power consumption, and still more if world population continues to increase, the exhaustion of fossil fuels would be even more rapid. The development of nuclear energy has improved this picture, but has not fundamentally altered it, at least in present technologies, for fissionable material is still relatively scarce. If we should achieve the economic use of energy through fusion, of course, a much larger source of energy materials would be available, which would expand the time horizons of supplementary energy input into an open social system by perhaps tens to hundreds of thousands of years. Failing this, however, the time is not very far distant, historically speaking, when man will once more have to retreat to his current energy input from the sun, even though this could be used much more effectively than in the past with increased knowledge. Up to now, certainly, we have not got very far with the technology of using current solar energy, but the possibility of substantial improvements in the future is certainly high. It may be, indeed, that the biological revolution which is just beginning will produce a solution to this problem, as we develop artificial organisms which are capable of much more efficient transformation of solar energy into easily available forms than any that we now have. As Richard Meier has suggested, we may run our machines in the future with methane-producing algae.[2]

The question of whether there is anything corresponding to entropy in the information system is a puzzling one, though of great interest. There are certainly many examples of social systems and cultures which have lost knowledge, especially in transition from one generation to the next, and in which the culture has therefore degenerated. One only has to look at the folk culture of Appalachian migrants to American cities to see a culture which started out as a fairly rich European folk culture in Elizabethan times and which seems to have lost both skills, adaptability, folk tales, songs, and almost everything that goes to make up richness and complexity in a culture, in the course of about ten generations. The American Indians on reservations provide another example of such degradation of the information and knowledge system. On the other hand, over a great part of

---

[2] Richard L. Meier, *Science and Economic Development* (New York: John Wiley and Sons, 1956).

human history, the growth of knowledge on the earth as a whole seems to have been almost continuous, even though there have been times of relatively slow growth and times of rapid growth. As it is knowledge of certain kinds that produces the growth of knowledge in general, we have here a very subtle and complicated system, and it is hard to put one's finger on the particular elements in a culture which make knowledge grow more or less rapidly, or even which make it decline. One of the great puzzles in this connection, for instance, is why the take-off into science, which represents an "acceleration," or an increase in the rate of growth of knowledge in European society in the sixteenth century, did not take place in China, which at that time (about 1600) was unquestionably ahead of Europe, and one would think even more ready for the breakthrough. This is perhaps the most crucial question in the theory of social development, yet we must confess that it is very little understood. Perhaps the most significant factor in this connection is the existence of "slack" in the culture, which permits a divergence from established patterns and activity which is not merely devoted to reproducing the existing society but is devoted to changing it. China was perhaps too well-organized and had too little slack in its society to produce the kind of acceleration which we find in the somewhat poorer and less well-organized but more diverse societies of Europe.

The closed earth of the future requires economic principles which are somewhat different from those of the open earth of the past. For the sake of picturesqueness, I am tempted to call the open economy the "cowboy economy," the cowboy being symbolic of the illimitable plains and also associated with reckless, exploitative, romantic, and violent behavior, which is characteristic of open societies. The closed economy of the future might similarly be called the "spaceman" economy, in which the earth has become a single spaceship, without unlimited reservoirs of anything, either for extraction or for pollution, and in which, therefore, man must find his place in a cyclical ecological system which is capable of continuous reproduction of material form even though it cannot escape having inputs of energy. The difference between the two types of economy becomes most apparent in the attitude towards consumption. In the cowboy economy, consumption is regarded as a good thing and production likewise; and the success of the economy is measured by the amount of the throughput from the "factors of production," a part of which, at any rate, is extracted from the reservoirs of raw materials and noneconomic objects, and another part of which is output into the reservoirs of pollution. If there are infinite reservoirs from which material can be obtained and into which effluvia can be deposited, then the throughput is at least a plausible measure of the success of the economy. The gross national product is a rough measure of this total throughput. It should be possible, however, to distinguish that part of the GNP which is derived from exhaustible and that which is derived from reproducible resources, as well as that part of consumption which represents effluvia and that which represents input into the productive system again. Nobody, as far as I know, has ever attempted to break down the GNP in this way, although it would be an interesting and extremely important exercise, which is unfortunately beyond the scope of this paper.

By contrast, in the spaceman economy, throughput is by no means a desideratum, and is indeed to be regarded as something to be minimized rather than maximized. The essential measure of the success of the economy is not production and consumption at all, but the nature, extent, quality, and complexity of the total capital stock, including in this the state of the human bodies and minds included in the system. In the spaceman economy, what we are primarily concerned with is stock maintenance, and any technological change which results in the maintenance of a given total stock with a lessened throughput (that is, less production and consumption) is clearly a gain. This idea that both production and consumption are bad things rather than good things is very strange to economists, who have been obsessed with the income-flow concepts to the exclusion, almost, of capital-stock concepts.

There are actually some very tricky and unsolved problems involved in the questions as to whether human welfare or well-being is to be regarded as a stock or a flow. Something of both these elements seems actually to be involved in it, and as far as I know there have been practically no studies directed towards identifying these two dimensions of human satisfaction. Is it, for instance, eating that is a good thing, or is it being well fed? Does economic welfare involve having nice clothes, fine houses, good equipment, and so on, or is it to be measured by the depreciation and the wearing out of these things? I am inclined myself to regard the stock concept as most fundamental, that is, to think of being well fed as more important than eating, and to think even of so-called services as essentially involving the restoration of a depleting psychic capital. Thus I have argued that we go to a concert in order to restore a psychic condition which might be called "just having gone to a concert," which, once established, tends to depreciate. When it depreciates beyond a certain point, we go to another concert in order to restore it. If it depreciates rapidly, we go to a lot of concerts; if it depreciates slowly, we go to few. On this view, similarly, we eat primarily to restore bodily homeostasis, that is, to maintain a condition of being well fed, and so on. On this view, there is nothing desirable in consumption at all. The less consumption we can maintain a given state with, the better off we are. If we had clothes that did not wear out, houses that did not depreciate, and even if we could maintain our bodily condition without eating, we would clearly be much better off.

It is this last consideration, perhaps, which makes one pause. Would we, for instance, really want an operation that would enable us to restore all our bodily tissues by intravenous feeding while we slept? Is there not, that is to say, a certain virtue in throughput itself, in activity itself, in production and consumption itself, in raising food and in eating it? It would certainly be rash to exclude this possibility. Further interesting problems are raised by the demand for variety. We certainly do not want a constant state to be maintained; we want fluctuations in the state. Otherwise there would be no demand for variety in food, for variety in scene, as in travel, for variety in social contact, and so on. The demand for variety can, of course, be costly, and sometimes it seems to be too costly to be tolerated or at least legitimated, as in the case of marital partners, where the maintenance of a homeostatic state in the family is usually regarded as much more desirable than the variety and excessive throughput of the libertine. There are problems here which the economics profession has neglected with astonishing singlemindedness. My own attempts to call attention to some of them, for instance, in two articles,[3] as far as I can judge, produced no response whatever; and economists continue to think and act as if production, consumption, throughput, and the GNP were the sufficient and adequate measure of economic success.

It may be said, of course, why worry about all this when the spaceman economy is still a good way off (at least beyond the lifetimes of any now living), so let us eat, drink, spend, extract and pollute, and be as merry as we can, and let posterity worry about the spaceship earth. It is always a little hard to find a convincing answer to the man who says, "What has posterity ever done for me?" and the conservationist has always had to fall back on rather vague ethical principles postulating identity of the individual with some human community or society which extends not only back into the past but forward into the future. Unless the individual identifies with some community of this kind, conservation is obviously "irrational." Why should we not maximize the welfare of this generation at the cost of posterity? *"Après nous, le déluge"* has been the motto of not insignificant numbers of human societies. The only answer to this, as far as I can see, is to point out that the welfare of the individual depends on the extent to which he can identify himself with others, and that the most satisfactory individual identity is that which identifies not only with a community in space but also with a community extending over time from the past into the future. If this kind of identity is recognized as desirable, then posterity has a voice, even if it does not have a vote; and in a sense, if its voice can influence votes, it has votes too. This whole problem is linked up with the much larger one of the determinants of the morale, legitimacy, and "nerve" of a society, and there is a great deal of historical evidence to suggest that a society which loses its identity with posterity and which loses

[3] K. E. Boulding, "The Consumption Concept in Economic Theory," *American Economic Review,* 35:2 (May 1945), pp. 1–14; and "Income or Welfare?," *Review of Economic Studies,* 17 (1949–50), pp. 77–86.

its positive image of the future loses also its capacity to deal with present problems, and soon falls apart.[4]

Even if we concede that posterity is relevant to our present problems, we still face the question of time-discounting and the closely related question of uncertainty-discounting. It is a well-known phenomenon that individuals discount the future, even in their own lives. The very existence of a positive rate of interest may be taken as at least strong supporting evidence of this hypothesis. If we discount our own future, it is certainly not unreasonable to discount posterity's future even more, even if we do give posterity a vote. If we discount this at 5 per cent per annum, posterity's vote or dollar halves every fourteen years as we look into the future, and after even a mere hundred years it is pretty small—only about 1½ cents on the dollar. If we add another 5 per cent for uncertainty, even the vote of our grandchildren reduces almost to insignificance. We can argue, of course, that the ethical thing to do is not to discount the future at all, that time-discounting is mainly the result of myopia and perspective, and hence is an illusion which the moral man should not tolerate. It is a very popular illusion, however, and one that must certainly be taken into consideration in the formulation of policies. It explains, perhaps, why conservationist policies almost have to be sold under some other excuse which seems more urgent, and why, indeed, necessities which are visualized as urgent, such as defense, always seem to hold priority over those which involve the future.

All these considerations add some credence to the point of view which says that we should not worry about the spaceman economy at all, and that we should just go on increasing the GNP and indeed the gross world product, or GWP, in the expectation that the problems of the future can be left to the future, that when scarcities arise, whether this is of raw materials or of pollutable reservoirs, the needs of the then present will determine the solutions of the then present, and there is no use giving ourselves ulcers by worrying about problems that we really do not have to solve. There is even high ethical authority for this point of view in the New Testament, which advocates that we should take no thought for tomorrow and let the dead bury their dead. There has always been something rather refreshing in the view that we should live like the birds, and perhaps posterity is for the birds in more senses than one; so perhaps we should all call it a day and go out and pollute something cheerfully. As an old taker of thought for the morrow, however, I cannot quite accept this solution; and I would argue, furthermore, that tomorrow is not only very close, but in many respects it is already here. The shadow of the future spaceship, indeed, is already falling over our spendthrift merriment. Oddly enough, it seems to be in

---

[4] Fred L. Polak, *The Image of the Future,* Vols. I and II, translated by Elise Boulding (New York: Sythoff, Leyden and Oceana, 1961).

pollution rather than in exhaustion that the problem is first becoming salient. Los Angeles has run out of air, Lake Erie has become a cesspool, the oceans are getting full of lead and DDT, and the atmosphere may become man's major problem in another generation, at the rate at which we are filling it up with gunk. It is, of course, true that at least on a microscale, things have been worse at times in the past. The cities of today, with all their foul air and polluted waterways, are probably not as bad as the filthy cities of the pretechnical age. Nevertheless, that fouling of the nest which has been typical of man's activity in the past on a local scale now seems to be extending to the whole world society; and one certainly cannot view with equanimity the present rate of pollution of any of the natural reservoirs, whether the atmosphere, the lakes, or even the oceans.

I would argue strongly also that our obsession with production and consumption to the exclusion of the "state" aspects of human welfare distorts the process of technological change in a most undesirable way. We are all familiar, of course, with the wastes involved in planned obsolescence, in competitive advertising, and in poor quality of consumer goods. These problems may not be so important as the "view with alarm" school indicates, and indeed the evidence at many points is conflicting. New materials especially seem to edge towards the side of improved durability, such as, for instance, neolite soles for footwear, nylon socks, wash and wear shirts, and so on. The case of household equipment and automobiles is a little less clear. Housing and building construction generally almost certainly has declined in durability since the Middle Ages, but this decline also reflects a change in tastes towards flexibility and fashion and a need for novelty, so that it is not easy to assess. What is clear is that no serious attempt had been made to assess the impact over the whole of economic life of changes in durability, that is, in the ratio of capital in the widest possible sense to income. I suspect that we have underestimated, even in our spendthrift society, the gains from increased durability, and that this might very well be one of the places where the price system needs correction through government-sponsored research and development. The problems which the spaceship earth is going to present, therefore, are not all in the future by any means, and a strong case can be made for paying much more attention to them in the present than we now do.

It may be complained that the considerations I have been putting forth relate only to the very long run, and they do not much concern our immediate problems. There may be some justice in this criticism, and my main excuse is that other writers have dealt adequately with the more immediate problems of deterioration in the quality of the environment. It is true, for instance, that many of the immediate problems of pollution of the atmosphere or of bodies of water arise because of the failure of the price system, and many of them could be solved by corrective taxation. If people had to pay the losses due to the nuisances which they create, a good deal more resources would go into the prevention of nuisances. These arguments involving external economies and diseconomies are familiar to economists, and there is no need to recapitulate them. The law of torts is quite inadequate to provide for the correction of the price system which is required, simply because where damages are widespread and their incidence on any particular person is small, the ordinary remedies of the civil law are quite inadequate and inappropriate. There needs, therefore, to be special legislation to cover these cases, and though such legislation seems hard to get in practice, mainly because of the widespread and small personal incidence of the injuries, the technical problems involved are not insuperable. If we were to adopt in principle a law for tax penalties for social damages, with an apparatus for making assessments under it, a very large proportion of current pollution and deterioration of the environment would be prevented. There are tricky problems of equity involved, particularly where old established nuisances create a kind of "right by purchase" to perpetuate themselves, but these are problems again which a few rather arbitrary decisions can bring to some kind of solution.

The problems which I have been raising in this paper are of larger scale and perhaps much harder to solve than the more practical and immediate problems of the above paragraph. Our success in dealing with the larger problems, however, is not unrelated to the development of skill in the solution of the more immediate and perhaps less difficult problems. One can hope, therefore, that as a succession of mounting crises, especially in pollution, arouse public opinion and mobilize support for the solution of the immediate problems, a learning process will be set in motion which will eventually lead to an appreciation of and perhaps solutions for the larger ones. My neglect of the immediate problems, therefore, is in no way intended to deny their importance, for unless we at least make a beginning on a process for solving the immediate problems we will not have much chance of solving the larger ones. On the other hand, it may also be true that a long-run vision, as it were, of the deep crisis which faces mankind may predispose people to taking more interest in the immediate problems and to devote more effort for their solution. This may sound like a rather modest optimism, but perhaps a modest optimism is better than no optimism at all.

Ezra J. Mishan: The Costs of Economic Growth

So far the critique of economic growth as a social priority has developed within a framework of basic assumptions. In particular, by accepting people's wants as something given to us independently of the workings of the economic system, it is possible to interpret its operation as tending to bring scarce resources into relation with people's wants. . . .

The most common of these basic assumptions, one frequently invoked to vindicate economic growth, is that any extension of the *effective* range of opportunities[1] facing a person (whether presented to him through the market or directly by the Government) contributes to an increase in his welfare. Similarly any reduction in the effective range of opportunities contributes to a diminution of his welfare.[2]

However, even in a market economy in which government intervention is at a minimum, there is one important opportunity that is denied to the customers; that of selecting the range of alternatives that will face him on the market. He can choose only from what is presented to him by the market—and a range of alternative physical environments is not the only thing that the market fails to provide. For one thing, the so-called exten-

From *The Cost of Economic Growth* by Edward J. Mishan. Frederick A. Praeger, Publishers, 1967. Reprinted by permission of the publisher.

[1] The word *effective* is inserted to indicate that the additional opportunities presented to him are relevant to his circumstances inasmuch as they induce him to select a new combination of goods and services in preference to the old combination which is, however, still available to him.

[2] This assumption, equating enlargement of effective choice with improved welfare, is closely connected with the assumption that the consumer knows his own interest best. This latter assumption is one which favours 'free choice' as against 'paternalism' in the distribution of goods. Insofar as government taxation (in order to provide goods or services) can be interpreted as a spending of people's money for them on goods or services that could be produced and/or distributed as economically through the market, one may legitimately talk of 'paternalism'. If the liberal economist ascribes a higher utility to a batch of goods that is freely chosen than to one of the same value that is, to some extent, prescribed by the Government he does so on the grounds that a man knows his own interests better than anyone else. The question of the empirical truth of this proposition may be held to be secondary to the political belief that we should act *as if* people did know their own interests best since actions based on other premises would be liable to result in undesirable social and political consequences. It is also possible, however, to regard the proposition as a *factual* assumption about human behaviour—a hypothesis that free choice always, or usually, or sooner or later, brings about a greater access of welfare to the individual than would be obtained under any system that restricted his freedom of choice.

sion of opportunities is not necessarily *effective,* in the sense defined. When new kinds of goods or new models of goods appear on the market the older goods or models are not always simultaneously available. They are withdrawn from production at the discretion of industry.

The argument purporting to show how consumers' wants ultimately control the output produced is facile enough: for it is, on the one hand, admittedly profitable to be first to discover and cater to a new want, while, on the other hand, it would seem unprofitable to withdraw from the market any good for which the demand continues undiminished. It would not be hard, therefore, to lay down conditions under which the wants of consumers tend quickly to influence the sorts of goods produced. But, unless the wants of consumers exist *independently* of the products created by industrial concerns it is not correct to speak of the market as acting to adapt the given resources of the economy to meet the material requirements of society. In fact, not only do producers determine the range of market goods from which consumers must take their choice, they also seek continuously to persuade consumers[3] to choose that which is being produced today and to 'unchoose' that which was being produced yesterday. Therefore to continue to regard the market, in an affluent and growing economy, as primarily a 'want-satisfying' mechanism is to close one's eyes to the more important fact, that it has become a want-*creating* mechanism.

This fact would be too obvious to mention, except that its implications are seldom faced. Over time, an unchanged pattern of wants would hardly suffice to absorb the rapid growth in the flow of consumer goods coming on to the markets of rich countries, the U.S. in particular, without the pressure afforded by sustained advertising.[4] In its absence, leisure, one suspects,

---

[3] Admittedly it is difficult in many circumstances to separate the informative from the persuasive elements of an advertisement, to say nothing of gauging the accuracy or relevance of the information provided. 'Smart people smoke Cancerettes!' is a claim which is not easy to test. If we defined the class of smart people we may discover that, in consequence of a prolonged and intensive advertising campaign, smart people have indeed taken to smoking Cancerette cigarettes even though they could not be distinguished from other brands when labels were removed. A picture of the product, or the name of the brand, printed without comment may well persuade people to buy more of the product. However, a case for the abolition of commercial advertising does not depend on such a distinction. Moreover, the abolition of commercial advertising cannot seriously be construed as an infringement of libertarian principle.

All that which is relevant in enabling the public to make a rational choice from the range of material goods and services offered by private enterprise may be more economically conveyed by an impartial body of analysts and administrators—an official or semi-official Consumers' Union in fact. One great argument in favour of this solution is the large saving in resources, both those expended by commerce (much of it in 'counter-advertising') and those wasted by the public as a result of unsatisfactory choices.

[4] The view that commercial advertising lowers the price of newspapers and journals deserves a comment. In the last resort the full economic costs of newspapers have to be borne by the public at large. But whether they end up paying for a large part of the newspaper through their purchases of the goods that are advertised, as at present,

would be increasing faster than it is. National resources continue to be used to create new wants. These new wants may be deemed imaginary or they may be alleged to be as 'real' as the original set of wants. What cannot be gainsaid, however, is that the foundation necessary to enable economists to infer and measure increases in individual or social welfare crumbles up in these circumstances. Only as given wants remain constant and productive activity serves to narrow the margin of discontent between appetites and their gratifications are we justified in talking of an increase of welfare. And one may reasonably conjecture that unremitting efforts directed towards stimulating aspirations and enlarging appetites may cause them to grow faster than the possibilities for their gratification, so increasing over time the margin of social discontent.

Be that as it may, in high consumption economies such as the United States, the trend is for more goods, including hardware, to become fashion goods. Manufacturers strive to create an atmosphere which simultaneously glorifies the 'pace-setter' and derides the fashion laggards. As productivity increases without a commensurate increase in leisure the accent shifts ever more stridently to boost consumption—not least to boost automobile sales although cities and suburbs are near-strangled with traffic—in order, apparently, to maintain output and employment. The economic order is accommodating itself to an indigestible flow of consumer gadgetry by inverting the rationale of its existence: 'scarce wants' have somehow to be created and brought into relation with rising industrial capacity.

Under such perverse conditions growthmen may continue, if they choose, to so juggle with words as to equate growth with 'enrichment', or 'civilization', or any other blessed word. But it is just not possible for the economist to establish a positive relationship between economic growth and social welfare. . . .

A final weakness of the link between expanding output and social welfare is revealed by consideration of what economists are wont to call 'the relative income hypothesis'—the hypothesis that what matters more to a person in a high consumption society is not his absolute real income, his

---

or whether, as an alternative possibility, they pay the full cost of the newspaper, free of commercial advertising while paying less for goods no longer advertised, is mainly a question of distribution. If we ignore the costs of real resources used in advertising —the services performed by advertising agencies, additional newsprint, etc.—it is purely a question of distribution; those people who buy more of the advertised products effectively subsidizing those newspaper readers who buy less of them. Once the real resources used up in advertising are brought into account, however, the public is paying more for both advertised goods and newspapers combined than it would pay for them in the absence of advertising.

The contribution of advertising, in terms of 'information' and 'entertainment' currently provided to the public, is not large relative to the resources used. Certainly the flow of relevant and impartial information could be multiplied and made available to the public for a fraction of the resources currently employed in the advertising industry.

command over material goods, but his position in the income structure of society. In its purest form, the thesis asserts that, given the choice, the high-consumption society citizen would choose, for example, a 5 per cent increase in his own income, all other incomes constant, to, say, a 25 per cent increase in his income as part of a 25 per cent increase in everybody's real income. The evidence in favour of the hypothesis in its purest form is not conclusive but it is far from being implausible, and in a more modified form it is hardly to be controverted. Our satisfaction with many objects depends upon their publicly recognized scarcity irrespective of their utility to us. It is not difficult to imagine the gratification experienced by a person living in a country in which all the other inhabitants are aware of his being the sole possessor of a radio, high-fi recorder, washing machine and other durables. Nor is it difficult to imagine his great satisfaction, arising from the knowledge of his being the sole possessor of these things, melting away as they become common household appurtenances; indeed, of his gradual dissatisfaction with them as he learns that his neighbours now possess far more advanced models than his own. However, the more truth there is in this relative income hypothesis—and one can hardly deny the increasing emphasis on status and income-position in the affluent society—the more futile as a means of increasing social welfare is the official policy of economic growth.

In sum, facile generalizations about the connection between expanding choice and social welfare which serve to quieten misgivings about the single-minded pursuit of economic growth are here rejected. The fact that what matters most to affluent-society man is not the increase of purchasing power *per se* but his relative status, his position in the income hierarchy, robs the policy of industrial growth of much of its conventional economic rationale. In part, this attitude of affluent-society man is to be explained by the central thesis of this volume: that beginning from the norms of post-war affluence economic growth has failed to provide men with additional choices significant to his welfare; that, indeed, it has incidentally destroyed some cardinal sources of welfare hitherto available. The bewildering assortment of gadgetry and fashion goods offers the sort of expansion that is as likely to subtract from than to add to his welfare. As producer, affluent-man has little choice but to adapt himself to the prevailing technology; no provision is made by industry enabling him, if he chooses, to forgo something in the way of earnings for more creative and enjoyable work. Nor, as citizen, has he yet been presented with the vital choice of quieter and more human environments, free of the ravages of unrestrained traffic.

Herman E. Daly: Toward a New Economics—Questioning Growth

> He looked upon us as a sort of animal to whose share, by what accident he could not conjecture, some small pittance of reason had fallen, whereof we made no other use than by its assistance to aggravate our natural corruptions and to acquire new ones which nature had not given us; that we disarmed ourselves of the few abilities she had bestowed, had been very successful in multiplying our original wants, and seemed to spend our whole lives in vain endeavors to supply them by our own inventions.
>
> Jonathan Swift

Any discussion of the relative merits of a stationary, no-growth economy, and its opposite, the economy in which wealth and population are growing, must recognize some important quantitative and qualitative differences between rich and poor countries and social classes. Consider the familiar ratio of gross national product (GNP) to total population (P). This ratio, per capita annual product (GNP/P), is the measure usually employed to distinguish rich from poor countries. In spite of its many shortcomings, it does have the virtue of reflecting in one ratio the two fundamental life processes of production and reproduction. Two questions must be asked of both numerator and denominator for both rich and poor nations: namely, what is the quantitative rate of growth, and qualitatively, exactly what is it that is growing?

The rate of growth in the denominator P is much higher in poor countries. While mortality is tending to equality at low levels throughout the world, fertility in poor nations is roughly *twice* that of rich nations. No other social or economic index divides the world so clearly and consistently into "developed" and "undeveloped" as does fertility.

Qualitatively, the incremental population in poor countries consists largely of hungry illiterates, while in rich countries it consists largely of well-fed members of the middle class. The incremental person in poor countries contributes negligibly to production, but makes few demands on world resources. The incremental person in a rich country adds to his country's GNP, but his high standard of living contributes greatly to depletion of the world's resources and pollution of its spaces.

The numerator, GNP, is growing at roughly the same rate in rich and poor countries—around 4 or 5 percent annually, with the poor countries probably growing slightly faster. Nevertheless, because of their more rapid population growth, the per capita income of poor countries is growing more slowly than that of rich countries. Consequently the gap between rich and poor widens over time.

Incremental GNP in rich and poor nations has very different qualitative significance. At some point, probably already passed in the United States, an extra unit of GNP costs more than it is worth. Extra GNP in a poor country, assuming it does not go mainly to the richest class of that country, represents satisfaction of relatively basic wants (food, clothing, shelter, basic education, etc.), while extra GNP in a rich country, assuming it does not go mainly to the poorest class, represents satisfaction of relatively trivial wants (more electric toothbrushes, yet another brand of cigarettes, more force-feeding through advertising, etc.).

The upshot of these differences is that for the poor, growth in GNP is probably still a good thing, while for the rich it is probably a bad thing. Growth in population, however, is a bad thing for both: for the rich because it makes growth in GNP less avoidable, and for the poor because it makes growth in GNP, and especially per capita GNP, more difficult. The following discussion is concerned exclusively with a rich, affluent-effluent economy such as that of the United States, and will seek to define more clearly the concept of a stationary-state economy, see why it is necessary, consider its economic and social implications, and finally, comment on an emerging political economy of finite wants and nongrowth.

The Art of Getting On

The term "stationary state" has had two quite different meanings in the history of economic thought. The original sense in which the classical economists used the term was that of an actual state of affairs toward which the real world was tending, a state in which growth in wealth and population will have ceased. The stationary state was the eschatology of political economy, the doctrine of the ultimate working out of the evolutionary forces of capitalism. The later meaning assigned to the term by the neo-classical school was that of an epistomologically useful fiction, an idealized abstraction like the frictionless machine of mechanics or the ideal gas of chemistry, which would serve as a stable analytical reference point in the study of the progressive or growing economy, but without any eschatological vision as to where the progressive economy would end. The latter sense is the one most current in economics today. However, it is the former, classical sense that is relevant to this discussion. Of course, adoption of the classical sense does not imply rejection of the neo-classical concept since the two are entirely different ideas.

This change in meaning was part of the general intellectual shift from teleology to mechanism. One cannot escape the impression that the neo-classical abstraction rose from servant to master as economists became more and more fascinated with working out all of the logical properties of this and other mechanistic abstractions and less concerned with the problems of the real world and just where "progress" was taking us. This drive towards logical purity and rigor in economic theory has cost a heavy price in terms of what Whitehead calls the "fallacy of misplaced concreteness," which "consists in neglecting the degree of abstraction involved when an actual entity is considered merely so far as it exemplifies certain [pre-selected] categories of thought."

The classical economists, on the other hand, were less rigorous, but not so prone to the fallacy of misplaced concreteness. Over a century ago John Stuart Mill, the great synthesizer of classical economics, spoke of the stationary state in words that could hardly be more pertinent today:

> It must always have been seen, more or less distinctly, by political economists, that the increase in wealth is not boundless: that at the end of what they term the progressive state lies the stationary state, that all progress in wealth is but a postponement of this, and that each step in advance is an approach to it.
>
> I cannot . . . regard the stationary state of capital and wealth with the unaffected aversion so generally manifested towards it by political economists of the old school. I am inclined to believe that it would be, on the whole, a very considerable improvement on our present condition. I confess I am not charmed with the ideal of life held out by those who think that the normal state of human beings is that of struggling to get on; that the trampling, crushing, elbowing, and treading on each other's heels which forms the existing type of social life, are the most desirable lot of human kind. . . . The northern and middle states of America are a specimen of this stage of civilization in very favorable circumstances . . . and all that these advantages seem to have yet done for them . . . is that the life of the whole of one sex is devoted to dollar-hunting, and of the other to breeding dollar-hunters. . . .
>
> I know not why it should be a matter of congratulation that persons who are already richer than anyone needs to be, should have doubled their means of consuming things which give little or no pleasure except as representative of wealth. . . . It is only in the backward countries of the world that increased production is still an important object: in those most advanced, what is economically needed is a better distribution, of which one indispensable means is a stricter restraint on population. The density of population necessary to enable mankind to obtain, in the greatest degree, all the advantages both of cooperation and of social intercourse, has, in all the most populous countries, been attained. . . . It is not good for a man to be kept perforce at all times in the presence of his species. . . . Nor is there much satisfaction in contemplating the world

> with nothing left to the spontaneous activity of nature. . . . If the earth must lose that great portion of its pleasantness which it owes to things that the unlimited increase of wealth and population would extirpate from it, for the mere purpose of enabling it to support a larger, but not a happier or better population, I sincerely hope, for the sake of posterity, that they will be content to be stationary, long before necessity compels them to it.
>
> It is scarcely necessary to remark that a stationary condition of capital and population implies no stationary state of human improvement. There would be as much scope as ever for all kinds of mental culture, and moral and social progress; as much room for improving the Art of Living and much more likelihood of it being improved, when minds cease to be engrossed by the art of getting on. Even the industrial arts might be as earnestly and as successfully cultivated, with this sole difference, that instead of serving no purpose but the increase of wealth, industrial improvements would produce their legitimate effect, that of abridging labor.

The direction in which political economy has evolved in the last hundred years is not along the path suggested in the quotation. In fact, most economists are hostile to the notion of stationary state and dismiss Mill's discussion as "strongly colored by his "social views" (as if the neo-classical theories were not so colored!); "nothing so much as a prolegomenon to Galbraith's *Affluent Society*"; or "hopelessly dated." The truth of the matter, however, is that Mill is even more relevant today than in his own time.

Discovering an Invisible Foot

Stationary state signifies a constant stock of physical wealth (capital), and a constant stock of people (population). Naturally these stocks do not remain constant by themselves. People die and wealth is physically consumed (worn out, depreciated). Therefore the stocks must be maintained by a rate of inflow (birth, production) equal to the rate of outflow (death, consumption). But this equality may obtain, and stocks remain constant, with a high rate of throughput (inflow equal to outflow) or with a low rate.

This definition of stationary state is not complete until the rates of throughput by which the constant stocks are maintained are specified. For a number of reasons the rate of throughput should be as low as possible. For an equilibrium stock the average age at "death" of its members is the reciprocal of the rate of throughput. The faster the water flows through the tank, the less time an average drop spends in the tank. For the population, a low rate of throughput (low birth and death rates) means a high life expectancy and is desirable for that reason alone—at least within limits.

For the stock of wealth, a low rate of throughput (low production and low consumption) means greater life expectancy or durability of goods and less time sacrificed to production. This means more "leisure" or non-job time to be divided into consumption time, personal and household maintenance time, culture time, and idleness. This too seems socially desirable.

But to these reasons for the desirability of a low rate of maintenance throughput, must be added some reasons for the impracticability of high rates. Since matter and energy cannot be created, production inputs must be taken from the environment, leading to depletion. Since matter and energy cannot be destroyed, an equal amount of matter and energy in the form of waste must be returned to the environment, leading to pollution. Hence lower rates of throughput lead to less depletion and pollution, higher rates to more. The limits regarding what rates of depletion and pollution are tolerable must be supplied by ecology. A definite limit to the size of maintenance flows of specific materials is set by ecological thresholds that, if exceeded, cause system breaks. To keep flows below these limits we can operate on two variables: the size of the stocks and the durability of the stocks. As long as we are well below these thresholds, economic cost-benefit calculations regarding depletion and pollution can be relied upon as a guide. But as these thresholds are approached, "marginal cost" and "marginal benefits" become meaningless, and Alfred Marshall's motto, "nature does not make jumps," and most of neo-classical marginalist economics becomes inapplicable. The "marginal" cost of one more step may be to fall into the precipice.

Of the two variables, size of stocks and durability of stocks, only the second requires further clarification. Durability here means more than just how long a commodity lasts. It also includes the number of times that the waste output can be reused as input in the production of something else. Nature has furnished the ideal model of a closed-loop system of material cycles powered by the sun. To the extent that our technology can imitate nature's solar-powered closed-loop, then our stock of wealth will tend to become as durable as our water, soil, and air which are the real sources of wealth since it is only through their agency that plants are able to capture vital solar energy. The ideal is that *all* physical outputs should be usable either as inputs in some other man-made process, or as non-disruptive inputs into natural material cycles.

The stationary state of wealth and population is maintained by an inflow of low entropy matter-energy and an outflow of an equal quantity of high entropy matter-energy. (Low entropy matter-energy is highly structured, organized matter and easily usable free energy. High entropy matter-energy is randomized, useless bits of matter, and latent, unusable energy.) Stocks of wealth and people feed on low entropy. Low entropy inputs are received from the environment in exchange for high entropy outputs to the environment. In this overall sense there can be no closed loop or recycling of both matter and energy because of the second law of thermodynamics.

However, within the overall system there can be subsystems of individual processes arranged so that their material input-output links form a closed loop. Conceivably all processes in the stationary state could be arranged to form a material closed loop. But the recycling of matter through this closed-loop "world engine" requires energy, part of which becomes irrevocably useless as it is dissipated into heat. Actually, industrial material cycles cannot be 100 percent closed as this would require an uneconomical, if not impossible expenditure of energy. Thus some of the high entropy output takes the form of randomized bits of matter, and some takes the form of heat. The limit to using energy to reduce material pollution is the resulting localized thermal pollution, not the very long run, universal thermodynamic heat death. Thus it is important to bear in mind that the expenditure of energy needed for recycling necessarily pollutes.

The mere expenditure of energy is not sufficient to close the material cycle, since energy must work through the agency of material implements. To recycle aluminum beer cans requires more trucks to collect the cans as well as more energy to run the trucks. More trucks require more steel, glass, etc., which require more iron ore and coal, which require still more trucks. This is the familiar web of inter-industry interdependence reflected in an input-output table.

All of these extra intermediate activities required to recycle beer cans involve some inevitable pollution as well. If we think of each industry as adding recycling to its production process, then this will generate a whole chain of direct and indirect demands on matter and energy resources that must be taken away from final demand uses and devoted to the intermediate activity of recycling. It will take more intermediate products and activities to support the same level of final output. The advantage of recycling is that it allows us to choose the least harmful combination of material and thermal pollution.

The classical economists thought that the stationary state would be made necessary by limits on the depletion side, but the main limits now seem to be in fact occurring on the pollution side. In effect, pollution provides another foundation for the economic law of increasing costs, but has received little attention in this regard since pollution costs are social while depletion costs are usually private. On the input side the environment is partitioned into spheres of private ownership. Depletion of the environment coincides, to some degree, with depletion of the owner's wealth, and inspires at least a minimum of stewardship. On the output side, however, the waste absorption capacity of the environment is not subject to partitioning and private ownership. Air and water are used freely by all and the result is a competitive, profligate exploitation—what biologist Garrett Hardin calls the "commons effect," what welfare economists call "external diseconomies," and what I like to call the "invisible foot."

Adam Smith's "invisible hand" leads private self-interest unwittingly to serve the common good. The "invisible foot" leads private self-interest to kick the common good to pieces. Private ownership and private use under a competitive market give rise to the invisible hand. Public ownership with unrestrained private use gives rise to the invisible foot. Public ownership with public restraint on use gives rise to the visible hand (and foot) of the planner. Depletion has been partially restrained by the invisible hand while pollution has been encouraged by the invisible foot. It is therefore not surprising to find limits occurring mainly on the pollution side.

Mini vs. Maxi

The economic and social implications of the stationary state are enormous and revolutionary. The physical flows of production and consumption must be *minimized, not maximized,* subject to some agreed upon minimum standard of living and population size. The central concept must be the stock of wealth, not as presently, the flow of income and consumption. (Kenneth Boulding has been making this point since 1949, but with no effect on his fellow economists.) Furthermore, the stock must not grow. The important issue of the stationary state will be distribution, not production. The argument that everyone should be happy as long as his absolute share of the wealth increases, regardless of his relative share, will no longer be available. The arguments justifying inequality in wealth as necessary for savings, investment, and growth will lose their force. With income flows kept low, the focus will be on the distribution of the stock of wealth, not on the distribution of income. Marginal productivity theories and "justifications" pertain only to flows and therefore are not available to explain or justify the distribution of stock ownership.

It is hard to see how ethical appeals to equal shares can be countered. Also, even though physical stocks remain constant, increased income in the form of leisure will result from continued technological improvements. How will it be distributed if not according to some ethical norm of equality? The stationary state would make fewer demands on our environmental resources, but much greater demands on our moral resources. In the past a good case could be made that leaning too heavily on scarce moral resources, rather than relying on abundant self-interest, was the road to serfdom. But in an age of rockets, hydrogen bombs, cybernetics, and genetic control, there is simply no substitute for moral resources and no alternative to relying on them, whether they prove sufficient or not.

With constant physical stocks, economic growth must be in nonphysical goods, particularly leisure. Taking the benefits of technological progress in the form of increased leisure is a reversal of the historical practice of taking the benefits mainly in the form of goods and has extensive social implications. In the past, economic development has increased the physical output of a day's work while the number of hours in a day has, of course, remained constant, with the result that the opportunity cost of a unit of time in terms of goods has risen. Time is worth more goods, and a good is worth less time. As time becomes more expensive in terms of goods, fewer activities are "worth the time." We become goods-rich and time-poor. Consequently we crowd more activities and more consumption into the same period of time in order to raise the return on non-work time to equalize it with the higher returns on work time, thereby maximizing the total returns to total time. This gives rise to what Staffan Linder has called the "harried leisure class."

Not only do we use work time more efficiently, but also personal consumption time, and we even try to be efficient in our sleep by attempting subconscious learning. Time-intensive activities (friendships, care of the aged and children, meditation and reflection) are sacrificed in favor of commodity-intensive activities (consumption). At some point people will feel rich enough to afford more time-intensive activities even at the higher price. But advertising, by constantly extolling the value of commodities, postpones this point.

From an ecological view, of course, this is exactly the reverse of what is called for. What is needed is a low relative price of time in terms of commodities. Then time-intensive activities will be substituted for material-intensive activities. To become less materialistic in our habits, we must raise the relative price of matter. Keeping physical stocks constant and using technology to increase leisure time will do just that. Thus a policy of non-material growth, or leisure-only growth, in addition to being necessary for keeping physical stocks constant, has the further beneficial effect of encouraging a more generous expenditure of time and a more careful use of physical goods. A higher relative price of material-intensive goods may, at first glance, be thought to encourage their production. But material goods require material inputs, so costs as well as revenues would increase, eliminating profit incentives to expand.

In the 1930's the late Bertrand Russell proposed a policy of leisure growth rather than commodity growth and viewed the unemployment question in terms of the distribution of leisure. The following words are from his essay, "In Praise of Idleness":

> Suppose that, at a given moment, a certain number of people are engaged in the manufacture of pins. They make as many pins as the world needs, working (say) eight hours a day. Someone makes an invention by which the same number of men can make twice as many pins as before. But the world does not need twice as many pins. Pins are already so cheap that hardly any more will be bought at a lower price. In a sensible world, everybody concerned in the manufacture of pins would take to working four hours instead of eight, and everything else would go on as before. But in the actual world this would be thought demoralizing. The men still work eight hours, there are too many pins, some employers go bankrupt, and half the men previously concerned in making pins are thrown out of

> work. There is, in the end, just as much leisure as on the other plan, but half the men are totally idle while half are still overworked. In this way it is insured that the unavoidable leisure shall cause misery all round instead of being a universal source of happiness. Can anything more insane be imagined?

In addition to this strategy of leisure-only growth, we can internalize some pollution costs by charging effluent taxes. Economic efficiency requires only that a price be placed on environmental amenities, it does not tell us who should pay the price. The producer may claim that the use of the environment to absorb waste products is a right that all organisms and firms must of necessity enjoy, and whoever wants air and water to be cleaner than it is at any given time should pay for it. Consumers may argue that the use of the environment as a source of clean inputs of air and water takes precedence over its use as a sink, and that whoever makes the environment dirtier than it otherwise would be should be the one to pay. Again the issue becomes basically one of distribution—not what the price should be, but who should pay it. The fact that the price takes the form of a tax automatically decides who should receive the price—the government. But this raises more distribution issues, and the solutions to these problems are ethical, not technical.

Another possibility of non-material growth is to redistribute wealth from the low marginal utility uses of the rich to the high marginal utility uses of the poor, thereby increasing total social utility. Joan Robinson has noted that this egalitarian implication of the law of diminishing marginal utility was "sterilized mainly by slipping from utility to physical output as the object to be maximized." As we move back from physical output to non-physical utility, the egalitarian implications become unsterilized.

Traditional Keynesian full employment policies will no longer be available to palliate the distribution question since they require growth. By allowing full employment, growth permits the old principles of distribution (income-through-jobs) to continue in effect. But with no growth in physical stocks and a policy of using technological progress to increase leisure, full-employment is no longer a workable principle of distribution. Furthermore, we add a new dimension to the distribution problem—how to distribute leisure.

A stationary population, with low birth and death rates, would imply a greater percentage of old people than in the present growing population, although hardly a geriatric society as some youth worshippers claim. Since old people do not work, the distribution problem is further accentuated. However, the percentage of children will diminish, so in effect there will be mainly a change in direction of transfer payments. More of the earnings of working adults will be transferred to the old and less to children.

What institutions will provide the control necessary to keep the stocks of wealth and people constant, with the minimum sacrifice of individual freedom? It would be far too simpleminded to blurt out "socialism" as the answer, since socialist states are as badly afflicted with growthmania as capitalist states. The Marxist eschatology of the classless society is based on the premise of complete abundance; consequently economic growth is exceedingly important in socialist theory and practice. And population growth, for the orthodox Marxist, cannot present problems under socialist institutions. This latter tenet has weakened a bit in recent years, but the first continues in full force. However, it is equally simpleminded to believe that our present big capital, big labor, big government, big military type of private profit capitalism is capable of the required foresight and restraint, and that the addition of a few effluent and severance taxes here and there will solve the problem. The issues are much deeper and inevitably impinge on the distribution of income and wealth.

Why do people produce junk and cajole other people into buying it? Not out of any innate love for junk or hatred of the environment, but simply in order to earn an income. If—with the prevailing distribution of wealth, income, and power—production governed by the profit motive results in the output of great amounts of noxious junk, then something is wrong with the distribution of wealth and power, the profit motive, or both. We need some principle of income distribution independent of and supplementary to the income-through-jobs link. A start in this direction was made by Oskar Lange, who attempted to combine some socialist principles of distribution with the allocative efficiency advantages of the market system. However, at least as much remains to be done here as remains to be done in designing institutions for stabilizing population. But before progress can be made on these issues we must recognize their necessity and blow the whistle on growthmania.

Stunting Growthmania

Although the ideas expressed by Mill have been totally dominated by growthmania, there are an increasing number of economists who have frankly expressed their disenchantment with the growth ideology. Arguments stressing ecological limits to wealth and population have been made by Kenneth Boulding and Joseph Spengler, both past presidents of the American Economic Association. Recently E. J. Mishan, Tibor Scitovsky, and Staffan Linder have made penetrating anti-growth arguments. There is also much in Galbraith that is anti-growth—at least against growth of commodities whose desirability must be manufactured along with the product.

In spite of these beginnings, most economists are still governed by the assumption of infinite wants, or the postulate of non-satiety as the mathematical economists call it. Any single want can be satisfied, but all wants in the aggregate cannot be. Wants are infinite in number if not in intensity, and the satisfaction of some wants stimulates others. If wants are infinite, growth is always justified—or so it would seem.

Even while accepting the above hypothesis, one could still object to growthmania on the grounds that given the completely inadequate definition of GNP, "growth" simply means the satisfaction of ever more trivial wants, while simultaneously creating ever more powerful externalities that destroy ever more basic environmental amenities. To defend ourselves against these externalities, we produce even more, and instead of subtracting the purely defensive expenditures, we add them. For example, the medical bills paid for treatment of cigarette-induced cancer and pollution-induced emphysema are added to GNP, when in a welfare sense they clearly should be subtracted. This should be labeled swelling, not growth. The satisfaction of wants created by brainwashing and "hogwashing" the public over the mass media also represents mostly swelling.

A policy of maximizing GNP is practically equivalent to a policy of maximizing depletion and pollution. This results from the fact that GNP measures the flow of a physical aggregate. Since matter and energy cannot be created, production is simply the transformation of raw material inputs extracted from the environment; consequently, maximizing the physical flow of production implies maximizing depletion. Since matter and energy cannot be destroyed, consumption is merely the transformation into waste of GNP, resulting in environmental pollution. One may hesitate to say "maximal" pollution on the grounds that the production inflow into the stock can be greater than the consumption outflow as long as the stock increases as it does in a growing economy.

To the extent that wealth becomes more durable, the production of waste can be kept low by expanding the stock. But is this in fact what happens? If one wants to maximize production, one must have a market. Increasing the durability of goods reduces the replacement demand. The faster things wear out, the greater can be the flow of production. To the extent that consumer reaction and weakening competition permit, there is every incentive to minimize durability. Planned obsolescence, programmed self-destruction, and other waste-making practices so well discussed by Vance Packard are the logical result of maximizing a marketed physical flow. If we must maximize something it should be the stock of wealth, not the flow—but with full awareness of the ecological limits that constrain this maximization.

But why this perverse emphasis on flows, this flow fetishism of standard economic theory? Again the underlying issue is distribution. There is no theoretical explanation, much less justification, for the distribution of the stock of wealth. It is a historical datum. But the distribution of the flow of income is at least partly explained by marginal productivity theory, which at times is even misinterpreted as a justification. Everyone gets a part of the flow—call it wages, interest, rent, or profit—and it all looks rather fair. But not everyone owns a piece of the stock, and that does not seem quite so fair. Looking only at the flow helps to avoid disturbing thoughts.

But even if wants were infinite, and even if we redefine GNP to eliminate swelling, infinite wants cannot be satisfied by maximizing physical production. As people grow richer they will want more leisure. Physical growth cannot produce leisure. As physical productivity increases, leisure can be produced by working fewer hours to produce the same physical output.

Even the common-sense argument for infinite wants—that the rich seem to enjoy their high consumption—cannot be generalized without committing the fallacy of composition. If all earned the same high income, a consumption limit occurs sooner than if only a minority had high incomes. The reason is that a large part of the consumption by plutocrats is consumption of personal and maintenance services rendered by the poor, which would not be available if everyone were rich. By hiring the poor to maintain and even purchase commodities for them, the rich devote their limited consumption time only to the most pleasurable aspects of consumption. The rich only ride their horses—they do not clean, comb, saddle, and feed them, nor do they clean the stables. If all did their own maintenance work, consumption would perforce be less. Time sets a limit.

The big difficulty with the infinite wants assumption, however, is pointed out by Keynes, who in spite of the use made of his theories in support of growth, was certainly no advocate of unlimited growth, as seen in the following quotation:

> Now it is true that the needs of human beings seem to be insatiable. But they fall into two classes—those needs which are absolute in the sense that we feel them whatever the situation of our fellow human beings may be, and those which are relative in the sense that we feel them only if their satisfaction lifts us above, makes us feel superior to, our fellows. Needs of the second class, those which satisfy the desire for superiority, may indeed be insatiable; for the higher the general level, the higher still they are. But this is not so true of the absolute needs—a point may soon be reached, much sooner perhaps than we are all of us aware of, when those needs are satisfied in the sense that we prefer to devote our further energies to non-economic purposes.

Lumping these two categories together and speaking of infinite wants in general can only muddy the waters. The same distinction is implicit in Mill, who spoke disparagingly of "consuming things which give little or no pleasure except as representative of wealth. . . ."

The source of growth lies in the use made of surplus. The controllers of surplus may be a priesthood that controls physical idols made from the surplus and used to extract more surplus in the form of offerings and tribute. Or they may be feudal lords, who through the power given by possession of the land extract a surplus in the form of rent and the corvée. Or they may be capitalists (state or private) who use the surplus in the form of capital to gain more surplus in the form of interest and quasi-rents.

If growth must cease, the surplus becomes less important and so do those who control it. If the surplus is not to lead to growth, then it must be consumed, and ethical demands for equal participation in the consumption of the surplus could not be countered by productivity arguments for inequality as necessary for accumulation. The surplus would eventually enter into the customary standard of living and cease to be recognized as a surplus. Accumulation in excess of depreciation, and the privileges attached thereto, would not exist.

We no longer speak of worshipping idols. Instead of idols we have an abomination called GNP, large parts of which, however, bear such revealing names as Apollo, Poseidon, and Zeus. Instead of worshipping the idol, we maximize it. The idol has become rather more abstract and conceptual and rather less concrete and material, while the mode of adoration has become technical rather than personal. But fundamentally, idolatry remains idolatry.

John Hardesty, Norris C. Clement, and Clinton E. Jencks: The Political Economy of Environmental Destruction

---

Man stands on the threshold of a new world. He also teeters on the brink of destruction, beset by the problems of war, poverty, racism, increasing probability of nuclear holocaust, potential world famine, and environmental destruction. Environmental destruction would appear to be the most certain of these threats. If man completely destroys his environment, it is 100 percent probable that he will in turn be destroyed. An increasing number of scientists are becoming convinced that in various crucial areas we may have already poisoned our environment beyond our capacity to restore it. We hear the cry for "zero population growth," but how often do we hear of the need for "zero GNP growth," particularly

---

This paper is an outgrowth of an earlier paper, "The Macroeconomics of Environmental Destruction" (unpublished), by John Hardesty. This is the first publication of this article.

from economists? In this essay we will seek to establish the necessity of a zero rate of growth of Gross National Product (GNP) in the United States, approximately determine the appropriate *level* of GNP, and examine the economic and social implications of such a radical change.

I. The Gross National Product–Environmental Destruction Linkage

Many people are familiar with the official National Income Accounts classification of (newly produced) national output—GNP—into consumer goods; consumer services; investment goods which include producers' plant and equipment, residential housing, and changes in business inventories; expenditures by government at all levels; and net exports. We would like to propose an alternative classification scheme which is more useful from the standpoint of environmental destruction. The purpose of the scheme is to reclassify the components of GNP according to their contribution (whether direct or indirect) to the destruction of the environment. Contrary to the National Income Accounts, which exclude double-counting, multiple counting of goods and services will be allowed in the following scheme in order to determine the significance of a particular product's link with environmental destruction. More than one entry for a particular good or service indicates more than one contribution to the environmental problem.

Virtually all goods and many of the services included in GNP can be placed in at least one of the following categories.

1. *Environment-destroying in consumption.* This category includes all those consumer goods with a significant effect on the environment occurring in the process of their consumption (use). Prime examples are the major contributor to air pollution, the automobile, and household detergent soaps.[1]

2. *Environment-destroying in production.* Included here are all those final goods and services produced in such a manner as to add to the environmental problem. Two subcategories are apparent.

a. There are those goods and services which directly involve production processes that pollute. Electricity production is an example inasmuch as coal-burning power companies rank second behind automobiles in contribution to air pollution and probably first in contribution of sulfur oxides, which frequently change chemically into airborne sulfuric acid.

---

[1] An argument can be made to include those foods and prescription drugs which do more harm than good to one's personal environment.

Production of electricity also contributes to water pollution. Steel, paper production, and air transportation are other obvious examples. Educational services are less obvious but are included nonetheless, since the capital goods involved (buildings, furnaces, etc.) are environment-destroying.

b. Included also are those final goods and services which cause pollution by using intermediate goods which themselves originate in pollution-linked production processes.[2] It is necessary to include this subcategory because GNP excludes intermediate goods (such as the paper used in making books), taking into account only those goods and services considered "final" products in order to avoid double-counting. From another point of view, if books need paper and paper production damages the environment, then a greater demand for books implies, other things being equal, greater damage to the environment. To obtain an idea of the scope involved here, notice that any product using steel as an intermediate good must be included, since steel mills are notorious air and water polluters. To the extent that farmers use pesticides and chemical fertilizers as intermediate goods, certain foods would also be included.

3. *Other environment-destroying services.* This category includes primarily services which are complementary to goods or other services that are pollution-linked. Examples would include repair and maintenance services such as those provided by auto mechanics.

4. *Environment-destroying in disposal.* Many of the nondurable goods we consume (particularly food and beverages) are packaged in nonreusable containers which are generally disposed of through burning, burying, or littering. Burning results in air pollution, burying in land pollution, and littering in water or land pollution, depending on the scene of the crime.

5. *Environment-destroying investments.* New investment here includes investment by government as well as producers' plant and equipment and residential housing.[3] Inventory increases, where appropriate, would be included in 1, 2, and 4 above and are not included in this category.

a. All producers' plant and equipment that directly pollute the environment are included here—for example, a coal-burning heating system for a manufacturing plant.

---

[2] We are inclined to include final goods and services produced by *capital* goods which themselves use such intermediate goods.

[3] Residential housing could, alternatively, be included in 2, above. Category 2 will also include most types of capital goods, depending on how they are produced.

b. Residential housing is included to the extent that it results in the destruction of agricultural land, forests, other natural resources, or necessary open space. It is important to recognize that suburban housing tracts are ordinarily only established after the nonmarket provision of an "adequate" means of transportation (usually a freeway); therefore, it is not simply a matter of the market valuing suburban housing more than agricultural land or natural resources. Of course there are also social costs (external diseconomies) involved in terms of psychological, and possibly biological, needs for open space and plenty of greenery.

c. Highways are an unambiguous example of an environment-destroying investment provided by government since they lead to noise and air pollution, cut up neighborhoods, and cause congestion.

This classification scheme, and the limited examples provided, give some indication of the nature and extent of the GNP–environmental destruction linkage. Very simply, the direct cause of the environmental problem is the conglomeration of goods, services, waste, junk, and implements of destruction called the Gross National Product—heretofore an agent of pride. For emphasis we repeat, many goods and services fit into the categorization in more than one place and are, therefore, "multi-destroyers" in the sense of contributing to destruction of the environment at each appearance.

The automobile, a prime example, is environment-destroying in consumption, production, and disposal and is probably the greatest single contributor to Gross National Product as well. New automobiles and parts cost the nation $37.2 billion in 1968, just slightly less than one half of all personal consumption expenditures for durable goods.[4] Gas and oil purchases accounted for another $19.0 billion. Auto repair and maintenance "labor" amounted to $6–7 billion. The subtotal is $62–63 billion or 11.5 percent of total consumption expenditures and 7.5 percent of GNP. Nor is this the entire story. A complete accounting must at least include the following:

1. Automobile-related investments in plant and equipment (including the purchase of trucks and automobiles for business purposes).
2. Services of automobile insurance companies.
3. Government purchases of new cars, trucks, parts, insurance, and the auto-related services of employees such as mechanics.
4. Government expenditures on highway construction and maintenance and motor vehicle departments.

---

[4] The source of all data in this section except the repair and maintenance figure is the *Economic Report of the President, 1969*. The repair and maintenance figure is estimated from a $9–10 billion estimated cost for all parts and labor.

5. Sixty percent of all expenditures to control air pollution and a large share of expenditures to combat urban congestion and decay.
6. A portion of the services of hospitals, doctors, police, mortuaries, life insurance companies, banks, credit unions, and finance companies.

All things considered, it would not appear unreasonable to attribute 10 percent of GNP directly (that is, aside from "multiplier" effects) to this single, most dangerous product.

Our point is that *given present technical knowledge and its application,* environmental destruction is primarily a function of the *level* of GNP, and the faster it grows the faster the environment is destroyed. While certain types of products contribute relatively more to environmental destruction, *most components of GNP are in some way environmental-destruction-linked.* For this reason changes in the composition of output alone are not sufficient as a long-run solution. This simply means that all developed countries, capitalist or socialist, must give up their unquestioning allegiance to the credo that "more is always better." If they do not, the desire for "more" will eventually outweigh the short-run gains made by marginal changes in output composition.

The concept of pollution taxes which internalize social costs of production, if applied on a large scale, would bring about both drastic changes in output composition and a lower rate of economic growth, perhaps even zero or negative. The negative effect on growth could be expected to occur for several reasons. First, the imposition of significant new taxes would tend to reduce consumer purchasing power, thus bringing about a decline in consumption expenditure and GNP. Neo-Keynesian economists would undoubtedly advocate compensatory fiscal and monetary policies in order to raise GNP back to its potential level. Our analysis suggests that such policies, if feasible, would lessen the beneficial impact of the pollution tax on the environment. Second, it is almost inconceivable that the aggregate demand for goods and services and GNP could be maintained at high levels (in the short run) because pollution taxes will fall most heavily on such mainstays of the economy as automobiles, petroleum, and defense. Liberals simply assume resources can be shifted to meet other (unlimited) needs. Alternative analyses, which seem more realistic to us, suggest that capitalism is an "irrational system" which must produce wasteful, destructive products in order to survive.[5] In other words, there are no adequate substitutes for autos, oil, and war in our capitalist economy. Third, high

---

[5] See especially Paul A. Baran and P. M. Sweezy, *Monopoly Capital* (New York: Monthly Review Press, 1966). For a concise confrontation of the liberal belief in the feasibility of conversion to a peacetime economy, see Michael Reich and David Finkelhor, "Capitalism and the 'Military-Industrial Complex': The Obstacles to Conversion," *Review of Radical Political Economics,* Vol. 2, No. 4 (Fall 1970).

pollution taxes would be imposed on capital goods, basic material inputs (for example, steel and paper), and fossil fuels due to their high degree of complicity in environmental destruction. Aside from short-run depressive effects, it is clear that a high level of economic growth could not be maintained in the long run, since it is precisely these sectors which provide the means for expansion.[6]

Economists generally do not care to examine the pollution tax scheme from this perspective. In fact, the concept is becoming increasingly popular because it is compatible with existing economic theory and appears to have the potential of combating environmental destruction with more flexibility than a flat zero growth restriction. The ecologists' lack of enthusiasm (they rarely mention pollution taxes) is not due, we believe, to their ignorance of the subject but to the unrealistic nature of the proposal. Pollution taxes are safely applicable only where there is virtual 100 percent knowledge of the environmental effects of a particular ecosystem disruption. Ecologists are well aware that such knowledge is currently inconceivable; the science of ecology is only in its infancy. As Murdoch and Connell admit, ecology is not more than 40 years old and much of it is still merely descriptive.[7] This simply means that we cannot determine the incidence, nature, or extent of the environmental destruction associated with existing products and production processes, nor can we make confident predictions for goods, services, and production methods yet to be developed.

What is significant about the pollution tax is its very high potential for becoming the environmental equivalent of the "War on Poverty" and "Vietnamization" hoaxes. In the main, these are not even token gestures; they are pseudo-events.[8] They give the impression that action is being taken to solve a particular problem when, in our opinion, this is not the case and was probably never intended to be.

So far in our discussion we have neglected possible changes in technology which might enable us to maintain the present composition and growth of output. Technological developments, at best, only reduce the average amount of environmental destruction associated with each dollar's worth of real GNP. It is unavoidable that continued increases in output will eventually overcome the effect of any technological improvement.[9] Blind faith

---

[6] In addition, pollution taxes would bear most heavily on low-income families. See Section VI below.

[7] William Murdoch and Joseph Connell, "All about Ecology," reprinted in Part 1 of this book.

[8] Murdoch and Connell point out that technological solutions also fall easily into the category of pseudo-events.

[9] Jay W. Forrester, M.I.T. "system dynamics" authority, has demonstrated this through the use of an elaborate computer simulation. Reduction of the "rate of pollution generation" by 50 percent merely delays the "pollution crisis" 20 years. His

in *continuous* future technological developments is foolish when action can be taken now to change man and his institutions. If such technological developments should occur in the future, they will constitute windfall gains in human welfare. Should they not work out, we will not suffer greatly. Let us examine a specific case in order to illustrate the point.

The energy to run our economy originates largely in the combustion process, that is, burning fuels. This combustion process pollutes the atmosphere with particulates, hydrocarbons, sulfur dioxide, nitrogen oxide, carbon monoxide, and carbon dioxide; it also consumes oxygen. Certainly we can conceive of possible technological solutions such as solar power and nuclear power (produced through fusion reaction, not nuclear fission, which creates significant pollutants of its own in the form of heat and radioactive waste); however, at this point, nobody is certain when, if ever, these power sources will become available for large-scale use, or even that some new and unheard of environmental problem will not be found to accompany their use.

To elaborate a bit further on this all-important point, let us define "entropy." Entropy is the potential of a system to perform work, and increasing entropy implies decreasing potential. The earth's entropy increases as work is performed, as the second law of thermodynamics states, and will continue to increase as "geological capital" is consumed. Although anti-entropic technology is conceivable, it is clearly not going to be available in the foreseeable future. Again, the risks of a blind faith in technological progress are enormous and unnecessary. As Kenneth Boulding, economist and environmentalist, says in *The Meaning of the Twentieth Century:*

> The results of failure [to develop anti-entropic technology], however, would be so momentous from the point of view of man and the evolution of this part of the universe that it would seem wise to make the most pessimistic assumptions possible, and for man at this stage to make a concerted and deliberate effort to avoid the waste of his exhaustible resources in war and luxury and to concentrate their use in expanding knowledge in the direction of achieving a closed-cycle, high-level system.[10]

---

solution: "an end to population and economic growth." See his "Counterintuitive Behavior of Social Systems," *Technology Review,* M.I.T., December 1970.

[10] Kenneth Boulding, *The Meaning of the Twentieth Century: The Great Transition* (New York: Harper & Row, 1964), p. 150. It is quite possible that the day will arrive when such technical knowledge will exist and understanding of the biosphere will be sufficiently complete to allow the introduction of non-destructive technology to proceed with rational economic growth, confident that there will be no unforeseen ecological side effects. That day is certainly far off, and the current problem is to see that it is possible to reach that day.

II. The Neo-Keynesian Growth Fetish

The United States is about to surpass a $1,000,000,000,000.00 (one trillion dollar) Gross National Product.[11] Even in terms of inflated prices this is certainly "gross" relative to conditions in the rest of the world, particularly to the approximately 2 billion people who suffer from malnutrition and undernourishment. This figure amounts to almost $5,000 per capita in current dollars, or $20,000 per family of four! Why, then, the economists' stress on growth when a more equal distribution of income could make "the affluent society" an unambiguous reality?

One explanation commonly given is that growth of GNP in real terms (i.e., adjusted for changes in the price level) is necessary to ensure full employment. This is only true, however, given capitalist institutions and the wage-labor definition of employment, and is only *necessary* given the "Puritan Ethic," which stipulates that he who does not work does not eat. Many economists refuse to acknowledge that acceptance of things as they are is no less subjective than advocacy of change.

Flowing directly from this acceptance of the status quo (or as they say, "the political realities") is the second popular rationalization. It is said that the growth of GNP, at fixed tax rates, will increase government tax revenues, thus providing funds which can be used to combat social and environmental problems without a redistribution of income. Unfortunately, acceptance of things as they are by economists and policymakers militates against the expenditure of large sums of money to bring about change. We have passed through a period of unprecedented economic growth in the 1960s, and we have yet to see an outpouring of funds for domestic purposes. The period of rapid growth and full employment has been in large part due to high levels of military spending. Neo-Keynesian economists, in general, have found this satisfactory.

It is clear, then, that the concept of zero GNP growth will initially find itself up against a wall of neo-Keynesian economists. It is also apparent that most radical economists have accepted the proposition that the welfare of any people depends on increasing production. Ultimately, as environmental reality asserts itself, it will be the inflexible economists who are up against the wall.

---

[11] The President's Council of Economic Advisers has predicted a $1.065 trillion GNP for 1971. More realistic estimates set the figure some $15–25 billion lower.

III. Population Growth and Environmental Destruction

Since environmental destruction is primarily a function of the level of real GNP, population increases must affect the environment through increments in GNP. It would seem reasonable to expect population growth to have an effect on such components of GNP as residential housing, automobiles, and food. In 1938 Alvin Hansen thought this connection was strong enough to entitle his Presidential Address to the American Economic Association "Economic Progress and Declining Population Growth" in response to the concurrence of economic stagnation and very slow population growth in the 1930s. Nevertheless, it must be noted that any component of GNP can increase or decrease regardless of the rate of population growth. Housing starts, for example, tend to be most responsive to changes in the rate of interest and the availability of credit. *To make population the significant variable it is necessary to assume a fixed standard of living.*

To put it another way, economists generally assume consumers' material wants are unlimited. Given the enormous quantity of resources allocated to advertising and a social system necessarily devoted to consumerism, the assumption is probably not unreasonable. While it is true that the typical middle-class family of four cannot readily use more than, say, three or four 150-horsepower automobiles, they may purchase new cars more often, and purchase automobiles with bigger engines and more gimmicks. Already some families are adding helicopters (which pollute three times more per passenger mile than automobiles) and small planes to the transportation stall, and, perhaps ultimately, each family may have a jet (supersonic?) or two all its own. The obvious point is, given the institutional structure, the population could remain constant (perhaps even decline) and GNP (and per capita consumption) would continue to grow—and along with it environmental destruction.

Empirically we find the 1960s one of the most prosperous periods in the history of this country, while at the same time exhibiting one of the lowest population growth rates. The population is currently growing by less than 2 million persons per year. In the light of this low, and evidently still declining, rate of growth of population, the goal of zero population growth takes on less urgency.[12] Indeed, we believe it can be safely concluded that

---

[12] This is not to mention the alarming genocidal and totalitarian undertones in some population control discussions. For example, one of our students suggested the solution to the poverty problem was to sterilize all the poor! For a slightly more sophisticated plan see Kenneth Boulding's "Green Stamp Plan" in his *The Meaning of the Twentieth Century*.

zero population growth is, in the short run and under prevailing conditions in the United States, neither necessary nor sufficient for a more rational treatment of the environment. On the other hand, zero GNP growth is a necessary and sufficient condition for preventing an increase in environmental destruction.[13] Let us look more closely at this concept.

IV. The Need for a Low-Level GNP

A static GNP at the current $1 trillion level will only keep environmental destruction from increasing. An actual reduction in GNP is called for if the detrimental effects of man's economic activities are to be minimized.

The production process adversely affects the environment in two ways. That is, we not only fill environmental sinks with waste but also use up huge quantities of irreplaceable natural resources. A recent article by biologist Wayne H. Davis (*The New Republic,* January 10, 1970) points out that the economies of the world could expect to grow at 4 percent annual rates indefinitely only if the world and all its resources could also expand by 4 percent per year. With zero GNP growth at the current level, many people might conclude that the United States would be doing its part to preserve natural resources. Unfortunately, this would not be so. The United States already consumes far more than its share of the minerals, fossil fuels, and other natural resources made available in the world each year. With approximately 6 percent of the world's population, the United States presently consumes anywhere from 35 to 50 percent of the annually available natural resources, and this may increase to as much as 60 to 80 percent of known stocks by the year 2000, according to Resources for the Future.[14] Taking the lowest figure of 35 percent and assuming a proportional relationship between percentage changes in GNP and percentage changes in natural resource inputs for convenience, GNP would have to be reduced to around $250 billion from a 1968 level of $865 billion in order to bring U. S. consumption of world resources down to the 10 percent level.[15] Notice that once the U. S. economy reaches this "bare bones" level and is not allowed to expand, zero population growth becomes an absolute necessity *in order to maintain a given standard of living.*

---

[13] Admittedly, even this statement oversimplifies reality by ignoring aspects of environmental destruction that are cumulative (i.e., a stock rather than a flow).

[14] As reported in Heather Dean, "Scarce Resources: The Dynamics of American Imperialism," Radical Education Project, Ann Arbor, Michigan.

[15] Note that 35 percent may well be too low, thus tending to offset the possible overreaction of GNP to changes in natural resource inputs assumed above.

V. The End of Economic Surplus

How could the people of the United States ever do without 60–70 percent of the GNP? The answer is quite simple. A very substantial portion of current GNP is either nonessential or is seriously detrimental to our lives. To the $80–90 billion[16] of automobile-linked expenditures already mentioned we can add most of the $78 billion spent on the military, $18 billion of advertising expenditures, more than $60 billion spent on net investment in plant and equipment (needed only for further growth), new middle-class and upper-class housing, and increases in business inventories, $15 billion for highways, $4 billion on the space program, and well over $6 billion for maintenance of the governmental bureaucracy. At this point our list of nonessential and environmentally destructive goods and services totals almost $275 billion, and we have hardly begun. We would have to include a substantial portion of the approximately $435 billion spent on consumer goods and services ("consumer crap").[17] Of this, about $235 billion went for basic necessities—food (including beverages), shoes, clothing, and shelter. Since gross farm product was less than $24 billion, it is clear that much of the $235 billion is payment for excess profits; the fancy supermarkets; the elaborate department stores; exquisite clothing shops located in enclosed, air-conditioned, landscaped shopping centers; junky, harmful drinks; and the nutrition-destroying techniques of food processors. In addition, a large share of the over $77 billion housing cost included in the $235 billion is imputed rental value (not actual expenditure) of owner-occupied homes.

Clearly, we are wasting irreplaceable human lifetimes and nonrenewable resources to produce hundreds of billions of dollars worth of unfulfilling trash which interrupts and destroys natural environmental processes. The data presented here are only suggestive, but even so there can be little doubt that a 50 percent or more reduction in GNP is realistic in terms of peoples' basic needs. Indeed, the actual cost of necessary personal consumption expenditures may be on the order of $100–150 billion.

Although not primarily concerned with the environmentally destructive characteristics of our economy, Professors Baran and Sweezy analyzed the generation and absorption of what they call *economic surplus*—the difference between what a society produces and the socially necessary costs of producing it. In *Monopoly Capital* statistical estimates of U. S. economic surplus are presented for each of the years 1929–1963. The long-term trend is slightly upward from 47 percent of GNP in 1929 to 56 percent in 1963.

---

[16] All data is for calendar 1968.

[17] Auto-related and advertising expenditures are not included in this figure.

It might be noted that a 60 percent economic surplus in 1968 implies necessary social costs equal to 40 percent of GNP, or about $340 billion—on the same order of magnitude suggested here for the low-level, static GNP.

VI. Zero GNP Growth and Poverty

The zero GNP growth position may, at first glance, appear elitist from the point of view of the American poor, who have not yet had the opportunity to experience and then reject materialism. That such a view might prevail is suggested by the tensions which sometimes exist between some neighboring "hippie" and minority communities. Nevertheless, a case can be made to show the compatibility of zero GNP growth with the aspirations of the poor.

Poor people in the United States are forced to suffer the effects of environmental destruction most directly. As workers, they must endure black lung disease (from coal dust) and pesticide-contaminated fields. Their vacations (escapes) are few and of limited duration. As consumers they cannot afford organically grown foods, a house in the suburbs, or a second home at Lake Tahoe. In other words, they are less able to pay to avoid the short-run effects of environmental destruction. If pollution taxes were to be imposed, poor and working-class people would pay proportionately more of this cost as well, particularly as corporations pass the tax along to the consumer. Higher product prices due to pollution taxes bear most heavily on those with lower incomes who must continue to purchase necessary commodities and thus pay disproportionately to solve a problem not of their own making.[18]

This question remains, however: would not the alternative of directly stopping economic growth do even more damage to the poor? We do not think so. The first point to note is the likelihood that even with the most optimistic economic growth assumptions, the year 1990 will see approximately 2.5 million poor families (not to mention some 5 million poor "unrelated individuals") in the United States.[19] Furthermore, virtually all of these families will be headed by women (without husbands) and will be almost completely impervious to the effects of further economic growth.

---

[18] For further discussion along these lines see Richard England, "The Distribution of Current Pollution Costs and of the Costs of Pollution Abatement," University of Michigan, 1970 (Mimeo.)

[19] See John Hardesty, *An Empirical Study of the Relationship between Poverty and Economic Prosperity* (unpublished Ph.D. dissertation), University of California, San Diego, 1970. In 1968 there were approximately 6.4 million poor families by the $3,000 (1963 prices) definition.

Probing deeper, we find that observed reductions in U. S. poverty are entirely the result of a fixed real income definition of poverty (for example, $3,000 in 1963 prices). Once it is recognized that poverty in an affluent society is primarily a *relative* phenomenon (which is not to deny the presence of hunger in the United States) and that in a society of millionaires the family receiving $20,000 per year still suffers material[20] and psychological privation, then the response of poverty to economic growth is quite different. In fact, Victor Fuchs finds the poor constituting approximately 20 percent of the population throughout the prosperous postwar period if poverty is defined as having an income less than 50 percent of the median (middle) family income.[21] This translates into an *increasing* absolute number of poor families, since total population has grown over the period. An argument can be made that the 50 percent is chosen arbitrarily; and so long as one family's (per capita) income is significantly less than another's, that family will feel (and be) deprived. The only necessary function of income inequalities is to serve as a work incentive in efficiency and growth-oriented societies (capitalist or socialist). Therefore, only a society practicing the zero GNP growth ethic will be sufficiently unconcerned with economic efficiency to allow equal per capita distribution of income, thus without question eliminating all income-defined poverty.

VII. Zero GNP Growth and the Third World

In addition to providing the basis for the release of human potential within the United States and reducing U. S. expropriation of the world's resources, there remains the question of possible effects on the people of poor countries. It is certainly not our intent to suggest that poor countries should stop trying to improve living standards. On the contrary, the policies suggested here would actually tend to raise incomes in poor countries and narrow the present gap between them and the developed areas.

First, it should be noted that the U. S. economy is vitally linked to those of poor countries via all aspects of international trade, aid, and private investment. In general, poor countries provide us with primary products (fuels, minerals, and agricultural products) while we supply them with capital (both monetary and physical) and other manufactured goods. Through our aid programs and private investments we also tend to link

---

[20] For example, it receives the least adequate shelter and medical care, and certainly its members have a lower life expectancy. In other words, what is viewed as "poverty" depends largely upon the level of living enjoyed by others.

[21] Victor R. Fuchs, "Toward a Theory of Poverty," in Chamber of Commerce of the United States, *The Concept of Poverty* (U. S. Government Printing Office: Washington, D. C.), 1965.

them to our technology (our way of thinking and doing things). Thus, the same methods and processes which have come to be recognized as environmentally destructive in our own country are being transmitted to the poor countries as fast as the international marketplace and national and international aid programs allow. To be sure, "Green Revolutions" and industrialization benefit some of the world's poor, but incalculable costs are involved which are not now being considered but which will have to be reckoned with someday.[22]

In the absence of international agreements and controls we could expect that pollution taxes imposed on U. S. producers and consumers in a capitalist economy would raise the prices of its products relative to those of competitors. This would then stimulate many U. S. corporations to establish their factories in other countries in order to escape the higher costs. The consequence of such a development would be to further aggravate the conflicts that now exist between rich and poor countries. First, the poor countries would suffer the lion's share of direct (local) environmental destruction which the United States would escape. In effect, then, the United States would become a relatively clean and well-kept "lawn" in the midst of the "cesspool" of the poor countries. Second, as U. S. firms moved to the poor countries, employment and incomes might increase, but so would U. S. profits and control over their economies. We could then expect to see our military obligations multiply in order to keep those investments safe from expropriations. All of this, of course, adds up to more imperialism and, in the long run, less independence and less improvement in the standards of living in the Third World.

There is little doubt that, under the conditions outlined above, environmental "cleanup" in the context of a capitalist economy would be inconsistent with long-run development in the Third World. Let us assume, now, that the zero GNP growth (at a low level) restriction were in effect in the United States. In what way would this policy affect the development of the poor countries?

The people of the United States could provide assistance equivalent to the amount formerly wasted on military expenditures and thereby transfer a part of our consumption and investment goods to the poor countries, implying a static GNP of perhaps $380 billion instead of $300 billion. Not only would this be a way of repaying a long overdue debt (after all, to a large extent, these countries financed the development of the now developed countries), but it would facilitate a more equal world distribution of necessary goods and services. Nationalization of foreign-based U. S. firms would also help repay that debt and abolish the flow of profits from the poor to the rich nations. In addition, the investment goods transferred to the poor countries would mean higher incomes and employment, which

---

[22] See, for example, *The New Alchemy Institute,* Spring 1971, pp. 1–15.

would more than offset the losses caused by the reduced U. S. demand for primary products.

It is more than symbolic that we suggest resource transfers from the United States to poor countries approximately equal to present military expenditures. We feel that were the United States to abolish its war machine and relinquish the role of world policeman, other superpowers would not feel threatened and a firm basis for worldwide disarmament could be established. Given the removal of imperialism and its military support in poor countries, we would expect an increasing number of nationalist and socialist revolutions to occur. And in those countries where legitimate forces exist to bring about development, it would come about in accordance with indigenous values, traditions, and aspirations. Furthermore, we feel that the obstacles presented by current U. S. imperialistic policies in poor countries are so deep and far-reaching that were the United States and other developed countries to merely leave them alone, their chances for development would immediately increase.

Although this might sound naive and simplistic, there is historical evidence for such a view. In a recent article,[23] John Gurley, a distinguished economist at Stanford University, enthusiastically lauds the "Mao version" of economic development. Not only has the GNP per capita grown at an annual rate of 4 percent since 1949 (a higher rate than most developing countries), but it has done so without the usual unequal distribution of income and *without large doses of foreign aid.* Furthermore, the economic development of China under Mao (which Gurley insists is little understood and badly interpreted in this country) has proceeded simultaneously with the development of a "new communist man"—who is cognizant of man's social relations and responsibilities. Perhaps these new men and women are China's version of the new men and women currently developing in the United States—people who know how to live *with* their environment.

VIII. The End of Capitalism?

It is worthwhile to ask how such drastic changes in the functioning of the U. S. economy can be accomplished. It is clear that the economic tools necessary to reduce GNP exist within the context of current institutions in the form of standard monetary and fiscal policies. In fact, recent years have seen an essentially zero GNP growth policy as taxes have been raised,

---

[23] John Gurley, "The New Man in the New China," in *The Center Magazine,* May–June, 1970, reprinted in *Review of Radical Political Economics,* Vol. 2, No. 4 (Fall 1970). Also see Gurley's "Capitalist and Maoist Economic Development," *Monthly Review,* February 1971.

government spending reduced, and tight monetary policies maintained in order to fight inflation—not environmental destruction. Restricting growth, for whatever ends, inevitably results in less need for manpower and thus increased unemployment. Under capitalist institutions, unemployment is "bad medicine" which can only be tolerated in small doses. Thus, we find the government's reaction to forecasts of a 6.5 percent unemployment rate by fall 1971 to be a planned increase in the military budget, which had declined slightly for two consecutive years in response to public pressure.[24]

A somewhat more unorthodox method of achieving zero GNP growth is suggested by Warren Johnson in his paper "The Guaranteed Income as an Environmental Measure." Johnson's thesis is that a guaranteed annual income (GAI) could be used to encourage people to desert traditional work and consumption patterns by opening up new opportunities and fostering new life styles. Proper use of the GAI over a transitional period would tend to weaken traditional labor force attachments and encourage workers to drop out. Given that a zero GNP growth economy would involve considerable unemployment in the conventional sense, it seems clear that guaranteed support in some form would be necessary at least as an end if not as a means. However, even a limited negative income tax with a forced-labor requirement (Nixon's Family Assistance Program) is meeting resistance. Apparently the requirements of zero GNP growth (at a low level) are incompatible with capitalist institutions. Several points support this argument.

1. In reducing GNP from close to $1 trillion to, say, $300–$400 billion composed essentially of basic necessities, the majority of existing firms would be eliminated, including most of the largest, which are typically in such concentrated industries as automobiles, petroleum, and defense. The existing holders of economic and political power would never allow this to happen if they could help it.

2. Given zero GNP growth (at a low level) and the associated guaranteed support, men and women would automatically be free from the oppression of a basic capitalist institution—the labor market. Who would work for the capitalists under these conditions? It is well known that hierarchical control of the work process as well as division and specialization of labor creates worker alienation. We believe this alienation is manifest in the apparent frustration and dissatisfaction which cause some blue-collar workers to beat peace marchers and support George Wallace. It would seem, for many if not most workers, that labor services are only provided because alternative means of securing an "adequate" income do not exist. The ruling class carefully cultivates high standards of what is adequate,[25] realizing that gross materialism not only aids absorption of

---

[24] *Newsweek,* November 30, 1970.

[25] For 1966–1967 the Bureau of Labor Statistics' "modest but adequate" annual family budget was $9,100. In current prices this would be well over $10,000.

economic surplus but also fills a personal void and serves to counterbalance alienation. When production, economic growth, and gross materialism are defamed and a reasonable level of support is guaranteed to all as a right, no offsets to alienation remain. The outcome would almost certainly be the disintegration of the "formal" labor force, leaving, of course, considerable work to be done in cooperative ways outside the corporate structure.

3. Competition (and perhaps some technological progress and innovation) under conditions of zero GNP growth would lead to the domination of industries by progressively fewer and larger firms. The tendency Marx recognized for wealth to concentrate in ever fewer hands would accelerate. It is difficult to conceive of capitalists collecting vast amounts of useless wealth—much as a rockhound collects pretty stones—since profits could not substantially be reinvested or spent on luxury consumption goods. Either being a capitalist in such a world would be pointless or the collected wealth and power would be utilized to do away with the zero growth restriction.[26]

4. In Schumpeterian terms, the "process of creative destruction," the essential feature of modern capitalism through which it renews itself by developing new markets, new products, and new techniques, would be brought to an abrupt halt with the end of economic growth. As Schumpeter predicts (for other reasons), with the demise of the process of creative destruction comes the end of capitalism.

For these reasons (and certainly others) the ruling capitalist class will hardly impose zero GNP growth upon itself and will certainly resist efforts to impose it from below.

IX. The Role of the Counterculture in an "Imposition from Below"

The fixed stock of key natural resources, the limited capacity of the environment to withstand man's destructive interference, and the spread of socialism and socialist revolutions among the world's poor nonwhite majority determine a future in which the basic necessities of life will have to be distributed equally among all people, necessitating a reduced share for the United States. The game is much closer to being zero-sum than any of us realized.[27]

Alienation and dissatisfaction are growing among the American people

---

[26] A shift to direct foreign investment is not likely to be significant for reasons discussed in Section VII above.

[27] A zero-sum game is one in which winnings and losings add algebraically to zero; that is, you can only increase your share at my expense.

—blacks, Chicanos, Puerto Ricans, Indians, white youth, women, blue- and white-collar workers. While the various groups tend to react to existing conditions differently, one clear trend is apparent—that of combating alienation and oppression by struggling to create liberated communities within the mother country. The trend was originally set by black people who, beginning with Malcolm X, recognized the reality of their separation and the power which could arise from it. Increasingly Chicanos and Puerto Ricans, to some extent Indians, and most recently white youth are taking this path. While movements for revolutionary change among women, professionals, white-collar workers, and the industrial proletariat are highly significant and compatible with this development, we would like to emphasize the unique consistency of revolutionary counterculture communities with the demands of the world's poor and the natural environment.

Why counterculture? Simply because in the United States less work, significantly reduced consumption, and communal living conditions[28] are demanded by Mother Nature and the Third World, as well as by our own alienation. If we begin the struggle for meaningful lives now in massive and growing numbers, the economy could begin to move toward zero (or negative) GNP growth. This means an easing of the pressure on the natural environment, a material contribution to the cause of equal world distribution of necessary goods and services, and the beginnings of a life worth living for all peoples.

The idea of revolutionary counterculture communities is of particular significance: they have begun to emerge in such places as Berkeley, Isla Vista, Madison, Ann Arbor, and Cambridge; and hints of more have appeared in most cities of at least moderate size and in some rural areas.[29] If this trend should continue and massive quantities of labor services, human capital, and consumer demand are withdrawn from the economy, continued GNP growth will eventually become impossible. This process will be accentuated by the "active" conflict (as opposed to the "passive" conflict we have been describing) between the revolutionary communities and the power structure. As the economy of the dominant society becomes weaker and the communities (both nonwhite and white) and allies (inter-

---

[28] Among other things, communal living reduces necessary consumption by taking advantage of economies of scale in consumption; for example, cars, refrigerators, washing machines, and electric lights can be shared.

[29] It is obvious that such communities cannot be truly liberated until the entire society is transformed. The heavy police pressure in black, brown, and white youth communities is indicative of the recognized threat they pose to the established order. To those who fear that such concentrations invite massive attack and extermination, we can only reply that certain political realities have kept the United States from waging all out genocidal war against Vietnamese, blacks, and browns in the recent past. Also note that significant numbers of residents in these communities do maintain ties with the dominant society (through parents, friends, relatives), and many others are nonrevolutionary students, faculty, hippie capitalists, and the remaining elderly.

nal and Third World) become stronger, the conflict should grow sharper, eventually reaching a climax. The scene can be expected to resemble guerrilla warfare in Third World countries as more people become politicized by the conflict. In both cases, the people struggle for greater freedom, self-determination, and life itself against a common enemy—U. S. capitalism and technocracy.[30]

Whether or not the revolutionary community strategy for liberation proves viable, certain generalizations about the nature of the struggle can be made. Most importantly, it cannot be just a struggle for political power but must at the same time embody a profound effort to build humane lives and liberating institutions. In other words, both the 100 percent political ("straight") activist and the apolitical cultural dropout are doomed to failure unless they can get together. What is really needed is serious, cultural-political revolutionaries. In "The Politics of Ecology" Barry Weisberg says: "The true origin of what has yet to become an authentic movement is in the People's Park episode, in militant actions against corporate despoilers (including sabotage) and in the private as well as public attempts to create ecologically sound lives."[31] Lest anyone think saving the environment will be a picnic, listen to Weisberg again:

> If the State of California would defend a parking lot with the life of one person and the shooting of another 150, imagine the cost of taking back a forest, preventing an offshore drilling rig from being placed, blocking the construction of a nuclear power plant or tampering with the power/communication/food/transport systems which make America grow. But the sooner this happens the better. The sooner the spirit of People's Park infuses every ecological action, the brighter will be our chances to insure the conditions for our survival and, beyond that, a decent society.[32]

X. A Vista of Freedom

What kind of world awaits us under zero GNP growth (at a low level)? As indicated, there will be substantial "unemployment" and/or "underemployment," thus divorcing both men and women from "work" and bringing about the need for a separation of basic material support from "employment." Quotation marks have been placed around "work," "em-

---

[30] Very simply, "technocracy" refers to high levels of bureaucracy, centralization, manipulation, efficiency, and impersonalization. For a discussion of youth's rejection of technocracy, see Theodore Roszak, *The Making of a Counter Culture* (New York: Doubleday & Co., 1969). Perhaps in order to stress his point Roszak goes too far in deemphasizing capitalism's role in the present low quality of life.

[31] *Liberation,* January 1970.

[32] Ibid.

ployment," "underemployment," and "unemployment" to emphasize the conventional usage of these words. In fact, the prospect of permanent "unemployment" is a prospect of freedom and unlimited opportunity for employment in service to mankind (for example, as "politically unattached" volunteers in underdeveloped countries) and in the pursuit of individual self-actualization—to be whatever one is capable of becoming, intellectually, in interpersonal relations, and in the development of artistic talents and manual skills.

If the cultural changes necessary to such a society have not taken place gradually during the pre-revolutionary period, then it will be necessary to institute coercive governmental controls to impose them on people. This, according to our view, would be extremely unfortunate and can be avoided by adhering to political-cultural strategies discussed above.

The dominant, unifying themes among the various segments of the movement to change America are self-determination and the desire to participate directly in the decisions which affect one's life. Assurance of this opportunity and minimization of potential abuse of power require that ultimate political power be vested in the individual communities themselves.

In these communities, relative lack of market-provided goods and services should further encourage the development of community feelings, cooperation, and nonmarket provision of services and handcrafted goods. Nonmarket provision implies supply without *quid pro quo*;[33] that is to say, entire communities might begin to act like extended families. The *market provision* of goods and services, that is, GNP, could be expected to decline, perhaps approaching zero after a considerable period of time. Note that this would require "the withering away of the state" and all the services it provides, as well as the taxes it collects. If people come together to provide consumer goods and services for themselves and others without a *quid pro quo,* it is reasonable to expect them to provide the collective goods normally supplied by government.

Voluntary cooperation between communities desiring a modicum of transportation and communication services as well as intercommunal transfer of some raw materials and foodstuffs available only in certain limited geographical areas is feasible without setting up formal organizations or bureaucracies. Grossly inefficient it is, but then efficiency is not a concern in such a society! Similarly, economic benefits provided by the technology of mass production would be sacrificed in favor of smaller-scale, ecologically sound technology (of the type being investigated by the New Alchemists, an organization of ecologists and other scientists) and a large degree of community self-sufficiency.[34] While this view of the future is not the

---

[33] "Something in return."

[34] See the various bulletins of the New Alchemy Institute, Box 432, Woods Hole, Massachusetts 02543, particularly the Spring 1971 issue.

only one possible, it is consistent with the requirements of self-determination and participation, nonalienating work, opportunities for self-actualization, equal distribution of income and wealth, development of community feeling and concern for others, a more equal sharing of the earth's resources, an end to U. S. exploitation of the world's poor, a reduced threat from the stockpiling of nuclear weapons, and, of course, an end to environmental destruction.

It should be recognized that the entire discussion concerning the implications of zero GNP growth involves an implicit assumption about the nature of man. We assume that, in general, people will react positively to freedom—that most of the undesirable characteristics we see are the result of social conditions and institutions which are oppressive—private property, racism, sexism, poverty, and inequality are obvious examples—and bring out the worst in man. In the argot of the times, people are in need of sufficient "space" to enable them to come together in a noncompetitive, supportive, mutually beneficial, loving manner. Zero GNP growth can provide that space.

XI. Summary

The theme of this paper has been to argue that considerations of the environment and natural resources require that optimum production be set at a low, constant level equal to a "socially necessary" level of consumption and that this could not take place under capitalist institutions. There is, however, less need for planning or any other form of economic control mechanism, including the market. The analysis presented here suggests that a considerably reduced GNP is feasible, noting that the basic necessities of good, nutritious food (which few receive now), utilitarian clothing, and protection from the elements would continue to be provided but without the frills. What will have to be done without is the grossly materialistic consumption and military spending both we and the U. S. economy have come to depend upon. Such a society, if it can be achieved, is incompatible with capitalist and technocratic institutions and implies greater freedom than man has ever known. A struggle to achieve zero GNP growth is coincident with the struggles for human rights, individual and racial self-determination, and socialism. The concept of zero GNP growth opens a new vista in revolutionary strategy.[35]

---

[35] Not only are apolitical cultural dropouts making a positive contribution, according to our analysis, but so also are the local citizens groups organizing to oppose economic and population growth in their areas. Examples of the latter are the Lesser San Diego and Lesser Seattle organizations. While such activities are merely reformist on a local level, we recognize that the whole (i.e., the national effect) is more than the sum of its parts.

Finally, it must again be emphasized that while the discussion of zero GNP growth may appear utopian, it most definitely is not.[36] It is based on the objective conditions of environmental destruction rather than adolescent daydreams. If the arguments presented in this paper are correct, zero GNP growth, and perhaps zero GNP, will be ultimately realized, either positively through cultural-political revolution or negatively through complete depletion and destruction of the environment.

Warren A. Johnson: The Guaranteed Income as an Environmental Measure

It is ironic that one of the major consequences of the almost continuous economic growth since World War II has been the steady deterioration of the environment. Our incomes and standard of living have increased significantly so that the private worlds of most Americans are now complete with expansive homes and gardens, full garages, kitchens, and closets. Nearly all urban areas have seen the construction of luxury apartments, country clubs, residential areas, marinas, shopping centers, and office buildings. While the private sector has prospered, the public sector has been neglected; and that part of our environment that is not privately owned has declined in quality, beset with air and water pollution, noise, crime, congested roads and freeways, and crowded recreation areas. Industrial agriculture has diminished the viability of rural life, forcing people to cities that are already so large that the quality of urban life has diminished too.

But if the result of sustained economic growth has not been altogether satisfying, it is still considered to be far better than the only alternatives that are generally visualized: unemployment, recession, and eventually depression. Economists are confident they understand what is necessary for continued prosperity and high employment—real economic growth of something like 4 percent per year. Such a rate of growth can be maintained, in theory, through judicious application of governmental spending and taxing power, supplemented by monetary controls. Politicians are sometimes reluctant to apply the measures that economists prescribe, as

[36] In "The End of Utopia" in *Five Lectures* (Boston: Beacon Press, 1970), Herbert Marcuse states that only schemes which violate natural laws should be considered utopian.

when President Johnson tried to finance simultaneously the Vietnam War and domestic social welfare programs without increasing taxes. And sometimes economists disagree on the stringency of economic controls to be applied, mainly how much unemployment is an acceptable price to pay for reducing inflation. But at no time is the basic necessity of economic growth disputed: growth adequate to maintain full employment without inflation.

Growth has generally been considered synonymous with progress, with health and vitality. Stability, on the other hand, has usually been associated with stagnation. The first qualms about the value of growth have been about population. Encouraged by calculations of how many years it would take to reach standing room only, the goal of stopping population growth has been rapidly accepted. But while ingenious mathematics has been applied to population growth, it is only recently that similar thoughts have been turned toward economic growth. It has been young people primarily, who, so lacking in economic responsibility and blissfully unremembering of the Great Depression, point out that what most economists look on as economic health is in itself death by obesity, sooner or later. The mathematics is inexorable; if all goes according to our best hopes and a steady 4 percent annual growth is maintained, this would mean a doubling in 17 years, a quadrupling in 34, and an eightfold increase in 51 years. This is a much faster rate of increase than population, which in the United States will double in around 70 years at present growth rates. Maintaining growth requires a feverish pace of resource extraction and use of energy, generates many types of pollution, and requires new technologies and continued urbanization, all of which create the specter of ecological armageddon. Both economic growth and population growth will have to be stopped sooner or later in order to achieve a stable balance with our environment. The only question is when.

For the most part, the fear of runaway inflation or depression—particularly the latter—is the motivating force in our economic policies. The scale of social breakdown, of the human suffering and humiliation during the Great Depression, is closely related to the unanimous feeling in the United States that it cannot be allowed to happen again, at all costs. So now the economists find themselves with undreamed-of power in the federal government, for they are the major tacticians in the struggle to maintain full employment without inflation. And since growth is necessary to achieve this objective, growth we shall have. All the major participants in the political process—the administration, business, and labor—agree on this, to the degree that there is no basic ideological difference between them, only skirmishes over how the end product, the wealth created by the economy, shall be divided among the participants. Underlying the whole process is the concept of economic responsibility, which has come to perform the same social function that religious faith played in medieval Europe: disobedience might call forth the wrath of God (runaway inflation or depression), while pious acts can be trusted to lead to heaven (a cornucopia of consumer goods and services). For the most part, environmental problems will be considered a necessary part of our way of life, to be tolerated in order to obtain what we are conditioned to believe is the good life. After all, nothing is all good, and surely ours is a better life than has ever been realized before, all things considered.

Historically, the vastness of our land has allowed us to exercise our propensity to escape from our problems, and our economic system has encouraged this behavior. Initially this meant abandoning eroded farms or logged-over forest land. Now it means leaving our central cities to decay, strangled by declining employment and tax revenues, by increasing welfare costs and crime, by anachronistic transportation systems, and by racial polarization. Out of sight, out of mind; most of the affluent move to the suburbs and the new resort communities, where life is easier and residents are free from taxation to support the high costs of the central city. There they are able to provide lavish private environments for themselves, occupying their time with golf, gardening, relaxing, and entertaining friends. These private worlds are where our loyalties are; the real world, the larger environment, is disowned. Environmental problems, especially the urban ones, are seen as too massive and pervasive to solve. There will be talk of strong regulation and some conscience money as well, both primarily to search for technological solutions which are much easier to apply than solutions that require social change. But our tendency toward privatism suggests that we will not make the massive commitments necessary to make our cities a source of pride and satisfaction.

The Basic Economic Problem

In order to keep everyone employed it is necessary to consume everything that the labor force can produce; consumption must equal production at full employment. If production exceeds consumption, goods pile up on shelves and in warehouses, and plants lay off labor. Workers without paychecks don't buy much, so sales fall off, which leads to other workers being laid off, and the whole thing can degenerate into a recession or depression. Recently we experienced the other side of the problem where, because of Vietnam War expenditures, not enough could be produced to meet all the demands, so prices for available goods were bid up by consumers, hence inflation. But as production of materials for Vietnam is reduced and returning GIs look for jobs, the persistent problem of unemployment has returned. In this situation, conservation measures that put people out of work, like shutting down polluting factories, stopping the construction of power plants, reducing the consumption of scarce fossil fuels or the production of environmentally damaging consumer goods—all of which put workers out of jobs—will have an increasingly hard time gaining acceptance. Only job-producing conservation measures, such as the construction of sewage treatment plants or the development and production of air pollution control devices will find the going easy. It should be noted that, within this framework, the GNP is useful primarily as a measure of the number of jobs the economy can generate rather than some abstract measure of human welfare.

Why is continued growth necessary? Why can't we just stay where we are? Because a good portion of our work force is employed through investment in new industrial plants, office buildings, freeways, schools, and houses. If these new facilities were not needed, a lot of workers would be unemployed. (Population growth is valuable in this context because it sustains demand for goods and services.) The second reason why growth is so essential is that the productivity of the average worker is increasing, primarily through automation. New products must be developed and consumption increased to keep pace with this increased productivity, or workers are laid off. So the government's role is to see to it that money is pumped into the economy when a recession threatens to make sure that consumption and employment are maintained. Unfortunately, economic activity usually requires an urban location to be competitive, uses resources and energy, and generates pollution, and additional problems are created in the process of using and disposing of the things produced. Most services of the private sector of the economy are also based on goods, in marketing, retailing, and servicing them.

An analogy can be made between our economy and a speeding train. It is an extraordinarily powerful and efficient train, and it is traveling very fast already. It works beautifully except for this odd characteristic of having to keep speeding up all the time. Already, for a number of would-be passengers the train is going too fast for them to get on, and a number of the riders are not enjoying the frantic trip the way they expected they would and want to get off or at least slow down, while just a few are beginning to wonder how fast the train will ultimately have to go, and whether it can get the increasing amounts of fuel, water, and air it needs to keep going. It is a finite world.

The objective, of course, is to control the machine so that it will not have to speed up all the time, to set it at whatever speed is most desirable and can be maintained, or even to operate several trains at different speeds for various life styles of different travelers.

Overproduction is the major source of the problem; we work too much, given our level of technology. This could theoretically be overcome without unemployment if everyone worked less and less as productivity increased and new investment declined. Unfortunately, this alternative is being rejected at present. The long decline in the average number of hours worked per week has bottomed out and has even started back up over 40 hours, primarily because workers are taking second jobs. Either there are still so many things people want to buy, or the problems of paying for what they have are so difficult, or they do not know what to do with the increased spare time that is as desirable as the extra income. Time is money, and to most people, free time isn't worth the money they could be making.

Galbraith has suggested an alternative to the overproduction of consumer goods and services—the reallocation of resources from private consumption to public endeavors, including improvement of the environment. Instead of more cars, public rapid transit systems would be built. Instead of the market-directed expansion of existing cities we would build new cities. Instead of more summer homes, there would be more parks and recreation areas. Instead of private services, public services would be expanded, in education, in aid to the disadvantaged, the sick, the mentally ill, and the aged. It is very logical, and federal funds will probably increase in some of these areas but certainly not at the expense of private consumption. Taxes, if anything, will also be reduced rather than raised because of our bias toward privatism. Tax money, for the most part, is seen as money down the drain, as money that goes mainly for someone else's welfare. Many public expenditures primarily benefit the affluent, although we are reluctant to admit it. Urban renewal, freeways, defense, and law enforcement all gain strong support from the broad middle class compared to public housing, public transportation, aid to the educationally handicapped, or urban parks. Most of us already have our own private parks around our houses and our own private transportation systems. Money spent on our own homes is not really spent but invested, and can be enjoyed while we own it, and then we can recoup our investment when we move. Unfortunately, we do not think of public facilities as our own, something to be proud of, cared for, and encouraged. Nowadays there is far more talk of tax revolts than higher taxes for the expanded government expenditures that would be necessary to redirect the growth of the economy in ways the market does not now support.

Establishment economists rightly discern this lack of public spirit among us while agreeing that there are deficiencies in the public environment that need attention and money. Their rebuttal to Galbraith's proposal, however, is that the only way to get the necessary government funds is to encourage rapid economic growth, to get the additional tax revenues that a higher GNP will generate. But this certainly has not worked in the last twenty years. The undesirable social and environmental effects of economic growth have grown much faster than tax revenues for financing solutions. Today we can barely provide funds for basic governmental functions such as education and welfare, let alone all the vast new needs for reconstructing decaying central cities, building new transportation systems, and acquiring parks and open space. Yet the GNP is now more than three times higher than it was twenty years ago. Instead of experiencing an abundance of tax revenues for the enhancement of the environment, we are losing ground rapidly. Is there any reason to believe the future will be otherwise?

Even if, for some unanticipated reason, we find ourselves with vast sums of public money, we may not be able to spend it effectively because measures to improve the environment impinge on so many private interests. Rapid transit threatens the auto and oil interests, public housing interferes with the private housing market, and acquiring land for parks or open space eliminates the developer and the craft unions. This is very different from spending for defense or space, where few if any economic interests are adversely effected while many benefit. What might be called modern environmental conservation, on the other hand, goes against much of the grain of our economic society; it requires strong governmental powers or heavy expenditures, the benefits are largely public rather than private, and it inhibits economic growth, the overriding objective of the system. These factors suggest that the deterioration of the environment will continue unless some way can be found to modify existing economic forces and to control the acceleration of the economy.

The Guaranteed Income

It would seem that some very basic changes are necessary. It is the objective here to suggest one, that of changing the nature of the work that our society permits and supports. At present, a prerequisite of almost all jobs is that they be economically "efficient," which generally means a high level of productivity and an urban location. Couldn't this prerequisite be relaxed? Couldn't we permit uneconomic forms of work in uneconomic locations? The classic situation is the small farmer who is being forced off the land or onto welfare roles by efficient industrial agriculture. Another would be the socially committed young person who cannot find a way to help the people who most need help and still subsist himself. Encouraging uneconomic forms of work in new ways may be the most straightforward method of slowing economic output, while broadening the range of opportunities for meaningful work at the same time. Significantly, the process that could lead to a true guaranteed income has begun in the form of President Nixon's Family Assistance Program. Its probability of coming into being seems fairly good, certainly as good, if not better, than the strong government regulations or expenditures that would be necessary to reverse the deterioration of the environment under the conditions of continued rapid growth.

For the most part, the guaranteed income has been considered as a welfare measure for those who are unable to work or who cannot find a job, and as a supplement to the incomes of the working poor. In the context of our environmental problems, however, the justification for the guaranteed income is that it offers a positive incentive for reduced production, for slowing the speed of our economic train. To consider the guaranteed income as a device for discouraging economic growth among the nonpoor majority of this country is in no way intended to denigrate the importance of the measure for those who live in poverty in our affluent society. Nor would the environmental effect be the same; among the poor it would almost certainly lead to increased work and consumption, the things the poor now need. But the effect of the guaranteed income on the poor is a subject in itself, one which is under intense study elsewhere. Here the subject is mainstream American society, where the economic machine has been fashioned that is taking us so rapidly in the direction of environmental deterioration.

In this context, a guaranteed income with a basic allowance of around $4,000 a year for a family of four could perform three important functions. It could (1) expand the opportunities for public service, (2) support the development of new methods of livelihood appropriate to an age where it is no longer necessary or even useful for everyone to have an economically efficient job, and (3) it could discourage economic growth while still providing a flexible device for maintaining the stability of our economic system.

Expanding Opportunities for Service

There is nothing more disheartening than to be asked by a young person what jobs are available to help with the social and environmental problems of our cities, rural areas, or reservations. For the most part, the opportunities for those with social concerns are in large, bureaucratic social welfare agencies where "professional" roles are prescribed, innovation is difficult, and heavy workloads lead to superficiality and coldness toward the poor, the aged, the unwanted children, the mentally disturbed. There are great needs in education, and there are many who want to teach, but not enough funds are available to put the two together. The guaranteed income would let students, ghetto residents, or anyone else with the motivation apply their energy and imagination to our social problems. Unstructured and flexible, it is the kind of situation that young people and ethnic groups have utilized effectively in recent years.

Fostering the Development of New Methods of Livelihood

The number of individuals looking for an opportunity for meaningful public service will not be large enough to get our accelerating economic machine under control. New methods of livelihood must also be developed. This trend is already well underway as the young especially, but others as well, attempt to escape the lonely and threatening world of competition and commercialism to find more satisfying ways. This is an uphill struggle, to a degree at least, because of the economic problems involved, unless special talents or outside financial resources are available. In the city, there are more diverse income possibilities, but costs of living are higher, public transportation is poor, and recreation is expensive or otherwise limited. The guaranteed income would almost certainly foster a movement back to rural areas because of the prospects for independence, offering inexpensive housing, the opportunity to raise at least part of one's food, and the use of wood, wind, or sun for energy. The small amounts of land needed for subsistence agriculture are relatively inexpensive. The guaranteed income would almost certainly contribute to the efforts of the Black Muslims and previously CORE to escape from the entrapment of urban ghettos. And if the surveys are correct that tell of large numbers of urban Americans who would like to return to small towns and the countryside, then the movement away from the cities would cover a broad spectrum of the urban population.

Probably the greatest problems for these emigrants will be their conditioning to fast-paced urban life and the loss of the rural skills of their forebears. It cannot be expected to be an easy transition, but at least they would have time to make mistakes and learn. The quality of the rural environment may initially decline as emigrants from urban areas learn new skills and new life styles evolve. The cultural (and environmental) shocks entailed in this process can at least be balanced against the avoidance of the cultural shock that occurs when people from depressed rural areas try to survive in the city without proper education or the support of kinfolk and neighbors. Both groups, however, those who stay on the land and those who move to it, would contribute to an easing of the problems of the cities.

One of the greatest benefits of the guaranteed income may very well be the encouragement it would offer for the reestablishment of community on a functional basis. With a low income, individuals would be encouraged to rely on community provision of some of their needs, as has been the case through most of the history of man. Community has declined recently because it serves no purpose when everyone has such a high standard of living that he can provide all of his needs on his own. But this privatism is what makes suburbia so uninteresting; there are few public places where people can work together, watch children play, or strike up a conversation. Suburban streets are quiet, save for a few children, and sidewalks are not even being built in some of the new developments. The result is that wives can't wait to get out and get a job, to see people, and to join the major source of action in our society. The guaranteed income would recreate the age-old conditions that fostered a viable and enriching community, a real need for cooperation and community provision of services and facilities.

Efforts to reinvigorate rural communities must be considered as valuable efforts to reestablish time-honored elements of civilization, a functioning community and a real interaction with nature. But there must be an infinite number of ways in which people for whom the city remains the center of gravity, could use a guaranteed income. Adoption of the guaranteed income would not have to be permanent, but could be an interlude in one's working career, before college or after, or an encouragement to leave a dull, dead-end job and train for a more satisfying one. Would-be craftsmen could quit their jobs and become working artisans. The guaranteed income would encourage labor intensive activity rather than the resource and energy intensive activities now encouraged. Each person, in considering the future, would have the opportunity to incorporate the guaranteed income in planning his life, balancing off the advantages of a job at high pay with the broadened opportunities of the guaranteed, but lower, income. Unlike President Nixon's present proposal, however, the guaranteed income would have to be available to anyone, not just those with a family who cannot get a job. If it were available to all, the guaranteed income would probably have better support from mainstream America, since everyone would be potential beneficiaries. At the same time, it would be less of a paternalistic welfare measure for the poor who are forced to depend on it. And jobs vacated by middle class workers might become available to the poor and unemployed, a most desirable effect.

Liberal education might also see a renaissance once the harsh demands of making a living were muted. Today, that query of "What will you do with a degree in art (or English or religions)?" has a real terror in it for students. Many of us have known a bright and sensitive student who wanted to be a poet or a musician but was not sufficiently confident in his abilities to take the brave step, realizing that the chances for making a living at it were almost nil. Because of the nature of jobs presently available and the competition for them, our educational system has become almost exclusively vocational and specialized, and the humanities have been virtually eclipsed at a time when they are needed more than ever. It could be expected that a whole new thrust in education would develop in response to the establishment of the guaranteed income, training in which the need for work would be balanced with other needs that are not now being met by our educational system.

The guaranteed income could function as a universal form of financial support for higher education. At present, scholarships are available to only a miniscule number of students, and loans are drying up, especially for undergraduates, while tuition and expenses are rising. A $1,000 guaranteed income would not solve all of the student's financial needs, but it would be a significant help.

In effect, what the guaranteed income would do would be to let people live the way they wanted to. For those who prefer the fast-paced, competitive, high-style life, this opportunity would be largely undiminished, and financial support would be available to obtain the education necessary to participate. For those who find this way of life unsatisfying, whatever the reason, the guaranteed income would provide the minimum resources to strike out in other directions. Hopefully, this would offer an incentive adequate enough to break our dependence on economic growth and put us on a track with greater evolutionary potential. It is frequently predicted that in the relatively near future only a fraction of the labor force will be needed to produce all of our basic needs. In such a situation why should everyone be forced into work that is economically productive but not socially or environmentally productive? Young people growing up in today's world are faced with what could be called economic tyranny. Well-paying jobs in large organizations are available, complete with all the elements generally referred to as the rat race. Other than something like the Peace Corps or VISTA, however, there are few other alternatives except dropping out completely. But economic tyranny being what it is, this is a very difficult alternative because food, medical care, and other needs are still expensive. At present, there is little middle ground; for the most part, you are either in the system or out of it, and for many individuals neither alternative is very satisfying. The guaranteed income would offer a major new alternative. Hopefully, as new methods of livelihood were developed, economic solutions demonstrated, and communities established, the impact of the guaranteed income would grow so that a number of ecological niches that are now vacant would be occupied.

A Flexible Device for Maintaining Economic Health without the Necessity of Growth

Even though the major long-run objective of the guaranteed income is to limit production and to remove the necessity of maintaining continuous economic growth, it is still absolutely essential that our economic system be maintained in a healthy state. Its productiveness is what permits the luxury of a guaranteed income in the first place (as well as creating the urgency for it). The requirement that total production of goods and services equals demand still holds, although now it is not at the previous level of full employment but at the lower level of the number of workers who wish to remain in economically productive jobs. But what happens if too many want to work, or too few, so that the desired amount of production is not maintained? The guaranteed income provides an effective device to help balance this equation; the government could raise or lower the amount of income guaranteed. For instance, if we should someday achieve a measure of peace which permitted dismantling a good portion of the military-industrial complex, the guaranteed income would be raised to encourage workers away from jobs, to take up the slack. Anytime unemployment became a problem, the guaranteed income could be raised, not only to attract workers but also to increase consumer expenditures. Thus, as productivity increases in the future, the guaranteed income would tend to rise, increasing its capacity to attract and absorb increased numbers of people just at the time when this would be necessary.

Lowering the guaranteed income, however, would cause problems, especially for people who have no other choice than the guaranteed income and who have families and bills for rent, food, and clothing. It is true that a reduction in the guaranteed income should be necessary only when there was a shortage of labor, so jobs would be available at the same time, but many recipients of the guaranteed income would be families without fathers, the incapacitated, and the elderly. A floor on the guaranteed income, limiting the amount it could be reduced to, could be a safeguard beyond which other government expenditures would have to be cut or taxes raised. Hopefully, the increased stability of an economy that does not have to grow continuously would reduce the economic fluctuations that would require changes in the guaranteed income. But still, the opportunity to vary the level of the income guaranteed would always be necessary, along with varying taxes and governmental spending, to maintain economic stability. People on a guaranteed income would have to accept the possibility of varying levels of income as a necessary part of life, in much the same way that the employed face varying job and income situations.

Prospects for a Guaranteed Annual Income

President Nixon's proposal for a Family Assistance Program is a significant step in the direction of a guaranteed income, even though the income provided is very low and is restricted to families with children and to those who can prove they cannot obtain a job. Opposition so far has been mainly from the Chamber of Commerce, and less openly from a few big city mayors who would prefer federal funds coming through the cities rather than directly to the recipients. It is also criticized by representatives of the poor who object to the low level of income provided and the coercive nature of the work and job training requirements. Opposition by conservatives has been limited, so far, mainly because Nixon's proposal anticipated this opposition and guarded against it while emphasizing the measure's definite advantages in providing incentives to work and to hold families together compared to present welfare programs.

However, if Nixon's proposal passed, and was subsequently expanded and liberalized into a full guaranteed income, it would then become a greater threat to traditional attitudes toward work and its rewards. The role of many welfare agencies would also be threatened, which would lead to the intense and uncompromising opposition that occurs whenever jobs are threatened. And the failures bound to occur with the first efforts to utilize the guaranteed income would receive full coverage in the media. It is impossible to anticipate what might happen during this critical stage. An optimistic scenario might include the difficulty of reducing the guaranteed income at this point because of the recession it would cause, or because middle class constituencies wanted to maintain it if unemployment intensified, or because state and local governments were dependent on federal financing and unwilling to support a more limited federal program which shifts responsibility back to state and local budgets. Key support would come from the large producers of consumer goods, whose main concern is not with a shortage of labor but with an adequate demand for their products; they would be the ones ending up with much of the guaranteed income anyway. The National Association of Manufacturers is now supporting the proposal for a Family Assistance Program.

Initially, the level of the guaranteed income would probably be determined more by the composition of the federal budget than by abrupt changes in taxes. The amount of money spent for such programs as defense and space, not the tax level, is the important factor. To keep the magnitude of the costs of a guaranteed income in perspective, consider that the 60 billion Department of Defense dollars (DOD's expenditures in 1970 were $77.8 billion) could provide 60 million people—with a guaranteed income of $4,000 for a family of four! A "peace scare," which now throws terror into millions of workers paid by DOD funds, could be utilized to slowly increase the guaranteed income while DOD funds were reduced. In the long run, however, after reducing the governmental expenditures that are largely to provide employment, it would seem that corporation taxes would be the logical source of revenue to finance the guaranteed income. As automation replaces workers, the new machines would be taxed to support the workers displaced, thus translating increased productivity into fewer economically productive jobs rather than increased consumption, as at present. The transition to the guaranteed income should be slow so that major social dislocations could be avoided and satisfactory ways to use the guaranteed income could evolve. The development of the guaranteed income should be slow, but if it was also a fitful and fluctuating development, anxieties and hardships would result. And it would be tragic if a preliminary program was terminated completely at a time of retrenchment and reaction, before it had time to stimulate a broad response.

There is no denying that there would be many problems. When the guaranteed income is first offered there might be disequilibrium in the labor market due to job resignations, but most of these jobs would be filled again when the restrictions of a low income were realized and other satisfying activities proved hard to find. If jobs with the lowest wages were widely vacated, especially the menial and repetitious, the pay for these jobs would have to be raised to get them done, to tempt workers back. Higher pay for such workers as agricultural laborers and janitors is only economic justice, but beyond that the upper level of the guaranteed income would have to be set carefully, and not be so high that work patterns would be altered excessively before social and economic conditions were ready. The more gradual the introduction of the guaranteed income the less would be the fluctuations in the labor market, although the impact would also be delayed.

One problem that demographers anticipate is an increase in the birth rates. Enjoying one's children is a pleasure that could be indulged in more fully on a guaranteed income; some people might wish to have children around much of the time, especially if they would mean increased income as well. Stabilizing the population, of course, is the other essential for survival in a livable world, along with stabilizing the economy, so positive steps would have to be taken to resist any trend toward increased birth rates. The obvious step would be to reduce the support for each additional child beyond the second to influence family planning. Perhaps, instead of each family's having more children, community enjoyment of children would develop, as in Europe, where townsfolk take great delight in the children playing in the streets, parks, and squares.

The most serious problem in the adoption of the guaranteed income would involve our real psychological dependence on work. For most of us, work is an outlet for the psychic energy generated by our active society, and because social status and rewards have been based on how productive one is, the ability to enjoy oneself has lost prestige, and its cultivation has declined. Work has had a tendency to become an end in itself, rather than something that supports other equally desirable elements of life. We have large voids in our lives which in older societies were filled with festivals, religious rites, folk arts, music, dancing, and just knowing how to relax. Older societies also had large amounts of essential work to round out their lives. It can be assumed that we need a certain measure of meaningful work, that without it we are troubled by alienation and boredom, by an existential vacuum. Man evolved as a working animal, to secure food, shelter, and warmth. To find satisfying work and to balance it with other elements of life may be a very difficult task. Let us hope that the opportunities for "uneconomic" work offered by the guaranteed income will stimulate fruitful results.

It is ironic that work, the one element we have depended on for most of our psychic income and have developed to a very high degree, is now becoming less useful and is even beginning to jeopardize our future. Our economic institutions and behavior are rapidly becoming obsolete and perilous. But since change in human behavior is difficult, it is important that the process of establishing a stable relationship with our environment be started as quickly as possible.

So often it seems that people have a tendency to underestimate the rate of change we are experiencing, failing to see that a very different future is being created. The rate of change today is unprecedented. Just a little over 50 years ago the Model T was introduced, and, as is commonly the case, only the benevolent aspects of the new technology were identified; Henry Ford saw it as a way to let Americans get out to God's green countryside. No one foresaw that the car would foster suburban sprawl that would bury the nearby countryside, lead to the decay of city life, consume vast quantities of resources, and become the major source of air pollution. Undesirable consequences are now appearing from the most humane technological advances; the control of disease in underdeveloped countries is leading to the far more terrible problems of unabated population growth and, inevitably, it seems, to death by starvation for many millions of people. For some time now it has been evident that it is the very nature of our technological achievements that is causing stress; it is because DDT is such an effective biocide, because machines can replace so much human labor, because the private car is such a flexible form of transportation, and because the concentration of economic activities in urban areas is so productive that our world is being disturbed now as never before. Most environmental measures now being considered have the uncomfortable air of being only stopgap measures, largely directed toward technological changes, which will have to be superseded by hoped-for technological breakthroughs, perhaps again and again as we continue to strain the limits of the environment, both physically and socially. The evolutionary potential of this line of action is definitely limited.

Yet the same technology provides us with vast power if it could be applied to support necessary social changes. The guaranteed income is one tool that seems to fit our social and environmental imbalances with remarkable directness, permitting economic behavior consistent with our technology but not tied desperately to it. It offers the possibility of balance and stability over the long run in our relation with the environment, if accompanied by a stabilized population.

In short, it has evolutionary potential.

E. F. Schumacher: Buddhist Economics

'Right Livelihood' is one of the requirements of the Buddha's Noble Eightfold Path. It is clear, therefore, that there must be such a thing as Buddhist Economics.

Buddhist countries, at the same time, have often stated that they wish

---

From *Resurgence,* Vol. 1, No. 11, Jan.–Feb. 1968. Reprinted by permission of the publisher.

to remain faithful to their heritage. So Burma: 'The New Burma sees no conflict between religious values and economic progress. Spiritual health and material well-being are not enemies: they are natural allies.'[1] Or: 'We can blend successfully the religious and spiritual values of our heritage with the benefits of modern technology.'[2] Or: 'We Burmans have a sacred duty to conform both our dreams and our acts to our faith. This we shall ever do.'[3]

All the same, such countries invariably assume that they can model their economic development plans in accordance with modern economics, and they call upon modern economists from so-called advanced countries to advise them, to formulate the policies to be pursued, and to construct the grand design for development, the Five-Year Plan or whatever it may be called. No one seems to think that a Buddhist way of life would call for Buddhist economics, just as the modern materialist way of life has brought forth modern economics.

Economists themselves, like most specialists, normally suffer from a kind of metaphysical blindness, assuming that theirs is a science of absolute and invariable truths, without any pre-suppositions. Some go as far as to claim that economic laws are as free from 'metaphysics' or 'values' as the law of gravitation. We need not, however, get involved in arguments of methodology. Instead, let us take some fundamentals and see what they look like when viewed by a modern economist and a Buddhist economist.

There is universal agreement that the fundamental source of wealth is human labour. Now, the modern economist has been brought up to consider 'labour' or work as little more than a necessary evil. From the point of view of the employer, it is in any case simply an item of cost, to be reduced to a minimum if it cannot be eliminated altogether, say, by automation. From the point of view of the workman, it is a 'disutility': to work is to make a sacrifice of one's leisure and comfort, and wages are a kind of compensation for the sacrifice. Hence the ideal from the point of view of the employer is to have output without employees, and the ideal from the point of view of the employee is to have income without employment.

The consequences of these attitudes both in theory and in practice are, of course, extremely far-reaching. If the ideal with regard to work is to get rid of it, every method that 'reduces the work load' is a good thing. The most potent method, short of automation, is the so-called 'division of labour' and the classical example is the pin factory eulogized in Adam Smith's *Wealth of Nations*. Here it is not a matter of ordinary specialization, which mankind has practised from time immemorial, but of dividing up every complete process of production into minute parts, so that the final

---

[1] *Pyidawtha, The New Burma.* (Economic and Social Board, Government of the Union of Burma, 1954, p. 10.)

[2] *Ibid.*, p. 8.

[3] *Ibid.*, p. 128.

product can be produced at great speed without anyone having had to contribute more than a totally insignificant and, in most cases, unskilled movement of his limbs.

Work

The Buddhist point of view takes the function of work to be at least threefold: to give a man a chance to utilize and develop his faculties; to enable him to overcome his ego-centredness by joining with other people in a common task; and to bring forth the goods and services needed for a becoming existence. Again, the consequences that flow from this view are endless. To organize work in such a manner that it becomes meaningless, boring, stultifying, or nerve-racking for the worker would be little short of criminal; it would indicate a greater concern with goods than with people, an evil lack of compassion and a soul-destroying degree of attachment to the most primitive side of this worldly existence. Equally, to strive for leisure as an alternative to work would be considered a complete misunderstanding of one of the basic truths of human existence, namely, that work and leisure are complementary parts of the same living process and cannot be separated without destroying the joy of work and the bliss of leisure.

From the Buddhist point of view, there are therefore two types of mechanization which must be clearly distinguished: one that enhances a man's skill and power and one that turns the work of man over to a mechanical slave, leaving man in a position of having to serve the slave. How to tell the one from the other? 'The craftsman himself,' says Ananda Coomaraswamy, a man equally competent to talk about the Modern West as the Ancient East, 'the craftsman himself can always, if allowed to, draw the delicate distinction between the machine and the tool. The carpet loom is a tool, a contrivance for holding warp threads at a stretch for the pile to be woven round them by the craftsman's fingers; but the power loom is a machine, and its significance as a destroyer of culture lies in the fact that it does the essentially human part of the work.'[4] It is clear, therefore, that Buddhist economics must be very different from the economics of modern materialism, since the Buddhist sees the essence of civilization not in a multiplication of wants but in the purification of human character. Character, at the same time, is formed primarily by a man's work. And work, properly conducted in conditions of human dignity and freedom, blesses those who do it and equally their products. The Indian philosopher and economist J. C. Kumarappa sums the matter up as follows.

---

[4] Ananda K. Coomaraswamy. *Art and Swadeshi.* (Ganesh and Co., Madras, p. 30.)

> If the nature of the work is properly appreciated and applied, it will stand in the same relation to the higher faculties as food is to the physical body. It nourishes and enlivens the higher man and urges him to produce the best he is capable of. It directs his freewill along the proper course and disciplines the animal in him into progressive channels. It furnishes an excellent background for man to display his scale of values and develop his personality.[5]

If a man has no chance of obtaining work he is in a desperate position, not simply because he lacks an income but because he lacks this nourishing and enlivening factor of disciplined work which nothing can replace. A modern economist may engage in highly sophisticated calculations on whether full employment 'pays' or whether it might be more 'economic' to run an economy at less than full employment so as to ensure a greater mobility of labour, a better stability of wages, and so forth. His fundamental criterion of success is simply the total quantity of goods produced during a given period of time. 'If the marginal urgency of goods is low,' says Professor Galbraith in *The Affluent Society,* 'then so is the urgency of employing the last man or the last million men in the labour force.' And again: 'If . . . we can afford some unemployment in the interest of stability—a proposition, incidentally, of impeccably conservative antecedents—then we can afford to give those who are unemployed the goods that enable them to sustain their accustomed standard of living.'[6]

From a Buddhist point of view, this is standing the truth on its head by considering goods as more important than people and consumption as more important than creative activity. It means shifting the emphasis from the worker to the product of work, that is, from the human to the subhuman, a surrender to the forces of evil. The very start of Buddhist economic planning would be a planning for full employment, and the primary purpose of this would in fact be employment for everyone who needs an 'outside' job: it would not be the maximization of employment nor the maximization of production. Women, on the whole, do not need an 'outside' job, and the large-scale employment of women in offices or factories would be considered a sign of serious economic failure. In particular, to let mothers of young children work in factories while the children run wild would be as uneconomic in the eyes of a Buddhist economist as the employment of a skilled worker as a soldier in the eyes of a modern economist.

While the materialist is mainly interested in goods, the Buddhist is mainly interested in liberation. But Buddhism is 'The Middle Way' and therefore in no way antagonistic to physical well-being. It is not wealth that

---

[5] J. C. Kumarappa. *Economy of Permanence.* (Sarva-Seva-Sangh-Publication, Rajghat, Kashi, 4th ed., 1958, p. 117.)

[6] J. K. Galbraith. *The Affluent Society.* (Penguin, 1962, pp. 272–273.)

stands in the way of liberation but the attachment to wealth; not the enjoyment of pleasurable things but the craving for them. The keynote of Buddhist economics, therefore, is simplicity and non-violence. From an economist's point of view, the marvel of the Buddhist way of life is the utter rationality of its pattern—amazingly small means leading to extraordinarily satisfactory results.

Standard of Living

For the modern economist this is very difficult to understand. He is used to measuring the 'standard of living' by the amount of annual consumption, assuming all the time that a man who consumes more is 'better off' than a man who consumes less. A Buddhist economist would consider this approach excessively irrational: since consumption is merely a means to human well-being, the aim should be to obtain the maximum of well-being with the minimum of consumption. Thus, if the purpose of clothing is a certain amount of temperature comfort and an attractive appearance, the task is to attain this purpose with the smallest possible effort, that is, with the smallest annual destruction of cloth and with the help of designs that involve the smallest possible input of toil. The less toil there is, the more time and strength is left for artistic creativity. It would be highly uneconomic, for instance, to go in for complicated tailoring, like the modern West, when a much more beautiful effect can be achieved by the skilful draping of uncut material. It would be the height of folly to make material so that it should wear out quickly and the height of barbarity to make anything ugly, shabby or mean. What has just been said about clothing applies equally to all other human requirements. The ownership and the consumption of goods is a means to an end, and Buddhist economics is the systematic study of how to attain given ends with the minimum means.

Modern economics, on the other hand, considers consumption to be the sole end and purpose of all economic activity, taking the factors of production—land, labour, and capital—as the means. The former, in short, tries to maximize human satisfactions by the optimal pattern of consumption, while the latter tries to maximize consumption by the optimal pattern of productive effort. It is easy to see that the effort needed to sustain a way of life which seeks to attain the optimal pattern of consumption is likely to be much smaller than the effort needed to sustain a drive for maximum consumption. We need not be surprised, therefore, that the pressure and strain of living is very much less in, say, Burma than it is in the United States, in spite of the fact that the amount of labour-saving machinery used in the former country is only a minute fraction of the amount used in the latter.

Pattern of Consumption

Simplicity and non-violence are obviously closely related. The optimal pattern of consumption, producing a high degree of human satisfaction by means of a relatively low rate of consumption, allows people to live without great pressure and strain and to fulfil the primary injunction of Buddhist teaching: 'Cease to do evil; try to do good.' As physical resources are everywhere limited, people satisfying their needs by means of a modest use of resources are obviously less likely to be at each other's throats than people depending upon a high rate of use. Equally, people who live in highly self-sufficient local communities are less likely to get involved in large-scale violence than people whose existence depends on world-wide systems of trade.

From the point of view of Buddhist economics, therefore, production from local resources for local needs is the most rational way of economic life, while dependence on imports from afar and the consequent need to produce for export to unknown and distant peoples is highly uneconomic and justifiable only in exceptional cases and on a small scale. Just as the modern economist would admit that a high rate of consumption of transport services between a man's home and his place of work signifies a misfortune and not a high standard of life, so the Buddhist economist would hold that to satisfy human wants from far-away sources rather than from sources nearby signifies failure rather than success. The former might take statistics showing an increase in the number of ton/miles per head of the population carried by a country's transport system as proof of economic progress, while to the latter—the Buddhist economist—the same statistics would indicate a highly undesirable deterioration in the *pattern* of consumption.

Natural Resources

Another striking difference between modern economics and Buddhist economics arises over the use of natural resources. Bertrand de Juvenal, the eminent French political philosopher, has characterized 'Western man' in words which may be taken as a fair description of the modern economist:

> He tends to count nothing as an expenditure, other than human effort; he does not seem to mind how much mineral matter he wastes and, far worse, how much living matter he destroys. He does not seem to realise at all that human life is a dependent part of an ecosystem of many different

> forms of life. As the world is ruled from towns where men are cut off from any form of life other than human, the feeling of belonging to an ecosystem is not revived. This results in a harsh and improvident treatment of things upon which we ultimately depend, such as water and trees.[7]

The teaching of the Buddha, on the other hand, enjoins a reverent and non-violent attitude not only to all sentient beings but also, with great emphasis, to trees. Every follower of the Buddha ought to plant a tree every few years and look after it until it is safely established, and the Buddhist economist can demonstrate without difficulty that the universal observance of this rule would result in a high rate of genuine economic development independent of any foreign aid. Much of the economic decay of South-East Asia (as of many other parts of the world) is undoubtedly due to a heedless and shameful neglect of trees.

Modern economics does not distinguish between renewable and non-renewable materials, as its very method is to equalize and quantify everything by means of a money price. Thus, taking various alternative fuels, like coal, oil, wood or water power: the only difference between them recognized by modern economics is relative cost per equivalent unit. The cheapest is automatically the one to be preferred, as to do otherwise would be irrational and 'uneconomic.' From a Buddhist point of view, of course, this will not do; the essential difference between non-renewable fuels like coal and oil on the one hand and renewable fuels like wood and water-power on the other cannot be simply over-looked. Non-renewable goods must be used only if they are indispensable, and then only with the greatest care and the most meticulous concern for conservation. To use them heedlessly or extravagantly is an act of violence, and while complete non-violence may not be attainable on this earth, there is none the less an ineluctable duty on man to aim at the ideal of non-violence in all he does.

Just as a modern European economist would not consider it a great economic achievement if all European art treasures were sold to America at attractive prices, so the Buddhist economist would insist that a population basing its economic life on non-renewable fuels is living parasitically, on capital instead of income. Such a way of life could have no permanence and could therefore be justified only as a purely temporary expedient. As the world's resources of non-renewable fuels—coal, oil and natural gas—are exceedingly unevenly distributed over the globe and undoubtedly limited in quantity, it is clear that their exploitation at an ever increasing rate is an act of violence against nature which must almost inevitably lead to violence between men.

---

[7] Richard B. Gregg. *A Philosophy of Indian Economic Development.* (Navajivan Publishing House, Ahmedabad, 1958, pp. 140–41.)

The Middle Way

This fact alone might give food for thought even to those people in Buddhist countries who care nothing for the religious and spiritual values of their heritage and ardently desire to embrace the materialism of modern economics at the fastest possible speed. Before they dismiss Buddhist economics as nothing better than a nostalgic dream, they might wish to consider whether the path of economic development outlined by modern economics is likely to lead them to places where they really want to be. Towards the end of his courageous book *The Challenge of Man's Future,* Professor Harrison Brown of the California Institute of Technology gives the following appraisal:

> Thus we see that, just as industrial society is fundamentally unstable and subject to reversion to agrarian existence, so within it the conditions which offer individual freedom are unstable in their ability to avoid the conditions which impose rigid organization and totalitarian control. Indeed, when we examine all of the foreseeable difficulties which threaten the survival of industrial civilization, it is difficult to see how the achievement of stability and the maintenance of individual liberty can be made compatible.[8]

Even if this were dismissed as a long-term view—and in the long term, as Keynes said, we are all dead—there is the immediate question of whether 'modernization,' as currently practised without regard to religious and spiritual values, is actually producing agreeable results. As far as the masses are concerned, the results appear to be disastrous—a collapse of the rural economy, a rising tide of unemployment in town and country, and the growth of a city proletariat without nourishment for either body or soul.

It is in the light of both immediate experience and long-term prospects that the study of Buddhist economics could be recommended even to those who believe that economic growth is more important than any spiritual or religious values. For it is not a question of choosing between 'modern growth' and 'traditional stagnation.' It is a question of finding the right path of development, the Middle Way between materialist heedlessness and traditionalist immobility, in short, of finding 'Right Livelihood.'

That this can be done is not in doubt. But it requires much more than blind imitation of the materialist way of life of the so-called advanced

---

[8] Harrison Brown. *The Challenge of Man's Future.* (The Viking Press, New York, 1954, p. 255.)

countries.[9] It requires above all, the conscious and systematic development of a 'Middle Way in technology,' as I have called it,[10, 11] A technology more productive and powerful than the decayed technology of the ancient East, but at the same time non-violent and immensely cheaper and simpler than the labour-saving technology of the modern West.

Harrison Brown: The Challenge of Man's Future

. . . The basic raw materials for the industries of the future will be seawater, air, ordinary rock, sedimentary deposits of limestone and phosphate rock, and sunlight. All the ingredients essential to a highly industrialized society are present in the combination of those substances. From seawater we will obtain water for agricultural and industrial purposes, hydrogen for iron-ore reduction, chlorine, caustic soda, magnesium, salt, bromine, iodine, and potassium, together with minor supplemental quantities of other elements. From the air we will obtain nitric acid and ammonia, which will be used for food production and industrial processing. From the air we will also obtain oxygen and other gases which are needed in various industrial operations. From ordinary rock we will satisfy the greater part of our requirements for essential metals and fissionable materials. From limestone we will derive the greater part of the carbon that will be a major starting point for the organic chemical industry and a starting point for the manufacture of liquid fuels. From phosphate rock we will obtain fertilizer. From sunlight we will obtain electricity and a variety of organic substances that will be used for food and as a starting point for various chemical industries.

The industries of the future will be far more complex and highly integrated than those of today. The "sea industries" will dwarf all existing mining operations. One can visualize vast assemblages of plants in coastal

[9] E. F. Schumacher. 'Rural Industries' in *India at Midpassage.* (Overseas Development Institute, London, 1964.)

[10] E. F. Schumacher. 'Industrialisation through Intermediate Technology' in *Minerals and Industries.* Vol. 1, no. 4. (Calcutta, 1964.)

[11] Vijay Chebbi and George McRobie. *Dynamics of District Development.* (SIET Institute, Hyderabad, 1964.)

regions where rock is quarried, uranium and other metals are isolated, nitric acid is manufactured, atomic power is generated, hydrogen is produced, iron ores are reduced to pig iron, aluminum and magnesium metals are prepared, and vast quantities of liquid fuels and organic chemicals are manufactured. As time goes on it is likely that the single-purpose plant will diminish in importance, eventually to disappear from the scene.

With increasing necessity and demand for efficiency, integration, and minimizing of waste in the economic world, there will be increasing demand for efficiency, integration, and minimizing of waste in the social world. These changes will have marked effects upon the ways in which men live. It seems clear that the first major penalty man will have to pay for his rapid consumption of the earth's non-renewable resources will be that of having to live in a world where his thoughts and actions are ever more strongly limited, where social organization has become all-pervasive, complex, and inflexible, and where the state completely dominates the actions of the individual.

. . . Within a period of time which is very short compared with the total span of human history, supplies of fossil fuels will almost certainly be exhausted. This loss will make man completely dependent upon waterpower, atomic energy, and solar energy—including that made available by burning vegetation—for driving his machines. There are no fundamental physical laws which prevent such a transition, and it is quite possible that society will be able to make the change smoothly. But it is a transition that will happen only once during the lifetime of the human species. We are quickly approaching the point where, if machine civilization should, because of some catastrophe, stop functioning, it will probably never again come into existence.

It is not difficult to see why this should be so if we compare the resources and procedures of the past with those of the present.

Our ancestors had available large resources of high-grade ores and fuels that could be processed by the most primitive technology—crystals of copper and pieces of coal that lay on the surface of the earth, easily mined iron, and petroleum in generous pools reached by shallow drilling. Now we must dig huge caverns and follow seams ever further underground, drill oil wells thousands of feet deep, many of them under the bed of the ocean, and find ways of extracting elements from the leanest of ores—procedures that are possible only because of our highly complex modern techniques, and practical only to an intricately mechanized culture which could not have been developed without the high-grade resources that are so rapidly vanishing.

As our dependence shifts to such resources as low-grade ores, rock, seawater, and the sun, the conversion of energy into useful work will require ever more intricate technical activity, which would be impossible in the absence of a variety of complex machines and their products—all of which are the result of our intricate industrial civilization, and which would be impossible without it. Thus, if a machine civilization were to stop functioning as the result of some catastrophe, it is difficult to see how man would again be able to start along the path of industrialization with the resources that would then be available to him. . . .

Our present industrialization, itself the result of a combination of no longer existent circumstances, is the only foundation on which it seems possible that a future civilization capable of utilizing the vast resources of energy now hidden in rocks and seawater, and unutilized in the sun, can be built. If this foundation is destroyed, in all probability the human race has "had it." Perhaps there is possible a sort of halfway station, in which retrogression stops short of a complete extinction of civilization, but even this is not pleasant to contemplate.

Once a machine civilization has been in operation for some time, the lives of the people within the society become dependent upon the machines. The vast interlocking industrial network provides them with food, vaccines, antibiotics, and hospitals. If such a population should suddenly be deprived of a substantial fraction of its machines and forced to revert to an agrarian society, the resultant havoc would be enormous. Indeed, it is quite possible that a society within which there has been little natural selection based upon disease resistance for several generations, a society in which the people have come to depend increasingly upon surgery for repairs during early life and where there is little natural selection operating among women, relative to the ability to bear children—such a society could easily become extinct in a relatively short time following the disruption of the machine network.

Should a great catastrophe strike mankind, the agrarian cultures which exist at the time will clearly stand the greatest chance of survival and will probably inherit the earth. Indeed, the less a given society has been influenced by machine civilization, the greater will be the probability of its survival. Although agrarian societies offer little security to the individual, they are nevertheless far more stable than industrial ones from a long-range point of view.

Is it possible to visualize a catastrophe of sufficient magnitude to obliterate industrial civilization? Here the answer must clearly be in the affirmative, for, in 1954, it takes no extraordinary imagination to foresee such a situation. Practically all major industrial countries are now aligned on one side or the other of a major dispute. Weapons of such power that whole cities can be destroyed in a few minutes are in the hands of the disputants, and, should a major war break out, those weapons, which become more powerful every year, will almost certainly be used. It is clearly within the realm of possibility that another war would so disrupt existing industrial societies that recovery would be impossible and the societies would either revert to agrarian cultures or become extinct. Indications of the possibilities that confront us are offered by the catastrophe which paralyzed Western Europe in World War II, and the slow process of its postwar recovery—a process which would have been very much slower had the highly industrialized United States not been in existence, relatively unscarred and prepared to give aid. And the damage and disruption of industrial activity we witnessed then are insignificant when compared with the disruption that might be suffered by all participants in an "atomic" war.

It is quite possible that a war fought at the present time, even with existing powerful weapons of mass destruction, would not bring industrial civilization to an end. With America and Europe prostrate, the people of Asia would have room into which they could expand and thus accelerate the evolution of their own industrial society. It is also quite possible that the West would recover from a major war, although admittedly recovery would be a far slower process than it was after World War II. But with each passing year, as populations become larger, as the industrial network becomes more complex, and as high-grade resources dwindle, recovery from a major war will become increasingly difficult.

It must be emphasized, however, that industrial civilization can come to an end even in the absence of a major catastrophe. Continuance of vigorous machine culture beyond another century or so is clearly dependent upon the development and utilization of atomic or solar power. If these sources of newly applied energy are to be available in time, the basic research and development must be pursued actively during the coming decades. And even if the knowledge is available soon enough, it is quite possible that the political and economic situation in the world at the time the new transition becomes necessary will be of such a nature that the transition will be effectively hindered. Time and again during the course of human history we have seen advance halted by unfavorable political and economic conditions. We have seen societies in which technical knowledge and resources were both present, but where adequate capital and organization were not in existence and could not be accumulated sufficiently rapidly. . . .

If industrial civilization eventually succumbs to the forces that are relentlessly operating to make its position more precarious, the world as a whole will probably revert to an agrarian existence. In such an event history will continue for as long a time as man exists. Empires, republics, and military states will rise and fall. There will be wars, migrations, and revolutions. Art, music, and literature will flourish, wane, then flourish again. As in the histories of the past and of the present, there will be unceasing change. Yet, looked upon over a period of thousands of years, history will have a sameness like the repeated performances of a series of elaborate epic plays in which, over the centuries, the actors change, the languages change, the scenery changes, but the basic plots remain invariant.

But if industrial civilization survives—if wars are eliminated, if the population of the world as a whole is stabilized within a framework of low death rates and low birth rates—will there continue to be a human history? The terms "stability" and "security" imply predictability, sameness, lack of change. And these terms further imply a high degree of organization—universal organization to avoid war, local organization to produce goods efficiently, and organization to control the distribution of goods. Organization in turn implies subjugation of the individual to the state, confinement and regimentation of the activities of the individual for the benefit of society as a whole.

Today we see about us on all sides a steady drift toward increased human organization. Governments are becoming more centralized and universal. In practically all areas of endeavor within industrial society—in our systems of production, in the fields of labor, capital, commerce, agriculture, science, education, and art—we see the emergence of new levels of organization designed to coordinate, integrate, bind, and regulate men's actions. The justifications for this increasing degree of organization to which man must accommodate himself are expressed in terms such as "stability," "security," and "efficiency." The end result of this rapid transition might well be the emergence of a universal, stable, efficient, industrial society within which, although all persons have complete personal security, their actions are completely controlled. Should that time arrive, society will have become static, devoid of movement, fixed and permanent. History will have stopped.

Here we indeed find ourselves on the horns of the dilemma. To what purpose is industrialization if we end up by replacing rigid confinement of man's actions by nature with rigid confinement of man's actions by man? To what purpose is industrialization if the price we pay for longer life, material possessions, and personal security in regimentation, controlled thoughts, and controlled actions? Would the lives of well-fed, wealthy, but regimented human robots be better than the lives of their malnourished, poverty-stricken ancestors? At least the latter could look forward to the unexpected happening—to events and situations which previously had been outside the realm of their experiences.

In a modern industrial society the road toward totalitarianism is unidirectional. In days gone by men could revolt against despotism. People could arise against their governments in the absence of legal recourse, and with muskets, sticks, knives, and stones as their weapons they could often defeat the military forces of the central authorities. But today our science and our technology have placed in the hands of rulers of nations weapons and tools of control, persuasion, and coercion of unprecedented power. We have reached the point where, once totalitarian power is seized in a highly industrialized society, successful revolt becomes practically impossible. Totalitarian power, once it is gained, can be perpetuated almost indefinitely in the absence of outside forces, and can lead to progressively more rapid robotization of the individual.

Thus we see that, just as industrial society is fundamentally unstable and subject to reversion to agrarian existence, so within it the conditions which offer individual freedom are unstable in their ability to avoid the conditions which impose rigid organization and totalitarian control. Indeed, when we examine all of the foreseeable difficulties which threaten the survival of industrial civilization, it is difficult to see how the achievement of stability and the maintenance of individual liberty can be made compatible.

The view is widely held in our society that the powers of the machine will eventually free man from the burden of eking out an existence and will provide him with leisure time for the development of his creativity and enjoyment of the fruits of his creative efforts. Pleasant though this prospect may be, it is clear that such a state cannot come into existence automatically; the pressures forcing man into devising more highly organized institutions are too great to permit it. If he is to attain such an idyllic existence for more than a transitory period he must plan for that existence carefully, and in particular he must do everything within his power to reduce the pressures that are forcing him to become more highly organized.

One of the major pressures that give rise to the need for increasing numbers of laws, more elaborate organization, and more centralized government is increase of population. Increase of numbers of people and of population density results in greater complexities in day-to-day living and in decreased opportunities for personal expression concerning the activities of government. But even more important, as populations increase and as they press more heavily upon the available resources there arises the need for increased efficiency, and more elaborate organizations are required to produce sufficient food, to extract the necessary raw materials, and to fabricate and distribute the finished products. In the future we can expect that the greater the population density of an industrial society becomes, the more elaborate will be its organizational structure and the more regimented will be its people.

A second pressure, not unrelated to the first, results from the centralization of industrial and agricultural activity and from regional specialization in various aspects of those activities. One region produces textiles, another produces coal, another automobiles, another corn, and another wheat. Mammoth factories require mammoth local organizations. Centralized industries must be connected, and this requires elaborate transportation systems. Regional localization of industries gives rise to gigantic cities, which in turn give rise to elaborate organization for the purpose of providing the inhabitants with the necessary food, water, and services. All of these factors combine to produce vulnerability to disruption from the outside, increased local organization and regimentation, more highly centralized government, and increasing vulnerability to the evolution of totalitarianism.

A third pressure results from increasing individual specialization and the resultant need for "integration," "coordination," and "direction" of activities in practically all spheres of vocational and leisure activity. It results in the placing of unwarranted trust in "integrators," "coordinators," and "directors." Early specialization results in lack of broad interests, lessened ability to engage in creative activity during leisure hours, decreased interest in the creative activities of other individuals, and lessened abilities to interpret events and make sound judgments. All of these factors combine to pave the way for collectivization, the emergence of strong organization, and, with it, the great leader.

Strong arguments can be presented to the effect that collectivization of humanity is inevitable, that the drift toward an ultimate state of automatism cannot be halted, that existing human values such as freedom, love, and conscience must eventually disappear.[1] Certainly if we used the present trends in industrial society as our major premises, the conclusion would appear to be inescapable. Yet is it not possible that human beings, recognizing this threat to the canons of humanism, can devise ways and means of escaping the danger and at the same time manage to preserve those features of industrial civilization which can contribute to a rich, full life? Is it really axiomatic that the present trends must continue and that in the long run industrial civilization and human values are incompatible? Here, in truth, we are confronted with the gravest and most difficult of all human problems, for it is one that cannot be solved by mathematics or by machines, nor can it even be precisely defined. Solutions, if they exist, can arise only in the hearts and minds of individual men.

The machine has divorced man from the world of nature to which he belongs, and in the process he has lost in large measure the powers of contemplation with which he was endowed. A prerequisite for the preservation of the canons of humanism is a reestablishment of organic roots with our natural environment and, related to it, the evolution of ways of life which encourage contemplation and the search for truth and knowledge. The flower and vegetable garden, green grass, the fireplace, the primeval forest with its wondrous assemblage of living things, the uninhabited hilltop where one can silently look at the stars and wonder—all of these things and many others are necessary for the fulfillment of man's psychological and spiritual needs. To be sure, they are of no "practical value" and are seemingly unrelated to man's pressing need for food and living space. But they are as necessary to the preservation of humanism as food is necessary to the preservation of human life.

I can imagine a world within which machines function solely for man's benefit, turning out those goods which are necessary for his well-being,

---

[1] These views have been forcefully and eloquently expressed by Roderick Seidenberg in his book *Post-Historic Man* (Durham: University of North Carolina Press, 1950).

relieving him of the necessity for heavy physical labor and dull, routine, meaningless activity. The world I imagine is one in which people are well fed, well clothed, and well housed. Man, in this world, lives in balance with his environment, nourished by nature in harmony with the myriads of other life forms that are beneficial to him. He treats his land wisely, halts erosion and overcropping, and returns all organic waste matter to the soil from which it sprung. He lives efficiently, yet minimizes artificiality. It is not an overcrowded world; people can, if they wish, isolate themselves in the silence of a mountaintop, or they can walk through primeval forests or across wooded plains. In the world of my imagination there is organization, but it is as decentralized as possible, compatible with the requirements for survival. There is a world government, but it exists solely for the purpose of preventing war and stabilizing population, and its powers are irrevocably restricted. The government exists for man rather than man for the government.

In the world of my imagination the various regions are self-sufficient, and the people are free to govern themselves as they choose and to establish their own cultural patterns. All people have a voice in the government, and individuals can move about when and where they please. It is a world where man's creativity is blended with the creativity of nature, and where a moderate degree of organization is blended with a moderate degree of anarchy.

Is such a world impossible of realization? Perhaps it is, but who among us can really say? At least if we try to create such a world there is a chance that we will succeed. But if we let the present trend continue it is all too clear that we will lose forever those qualities of mind and spirit which distinguish the human being from the automaton. . . .

Douglas Ross and Harold Wolman: Congress and Pollution—The Gentleman's Agreement

"When it comes to the environment," observed a conservationist Senator not long ago, "I'd guess you have to say that we have an informal understanding to talk big and carry a little stick."

Reprinted from *The Washington Monthly,* September 1970, by permission of the authors and the publisher.

He continued:

> Everyone in Congress realizes that it's a good issue to talk about back home because everyone there is for it. But those of us who know enough to be critical also know that the public is not ready to pay the price to really restore the environment. And nobody here wants to destroy the myth that Congress is militant on this issue: it would ruin a good thing for the Congressman and his colleagues.

Other new issues of the past few years—civil rights, Congressional control of foreign policy, the level of military spending, and the like—have produced the bitter disputes in the halls of the Capitol, the political divisions and alliances, and the floor fights over the final form of bills which mark these issues as genuine legislative concerns. To date, however, the fulmination over the environment has generated little conflict or debate, save for the feudal and somewhat carnival jousting by committees and their chairmen to determine who will win jurisdiction over this politically popular domain. In fact, this subject, which has been dubbed the "issue of the decade" by *Time, Newsweek,* and countless television documentaries—the crusade which many regard as the last and best hope for consensus in a strife-torn country and which some have gone so far as to see as the potential source for the transformation of American politics—falls under a kind of bipartisan "gentleman's agreement" in the Congress. The basic tenet of the agreement is that the ecology issue should remain a preserve for the rhetorical flourish which draws bountiful press coverage but which poses few political risks.

The environment is not yet even a political issue—the kind of problem which finds its way into election campaigns. Bussing, crime, and the Vietnam war are providing the focus in many Congressional elections, while a quick survey of the major races across the country must discourage those who look for anti-pollution politics to restructure the economic system. There is only one important election, the governor's race in Hawaii, where different views on pollution rank as a significant issue between candidates. Virtually no one is being attacked as soft, or hard, on the environment.

Finally, despite the enactment of six major water and air pollution control acts over the past decade—all by unanimous votes, or with overwhelming majorities—the quality of our air and water continues to deteriorate. In short, on those occasions when environmental concern has been translated into legislative action, the results have been inadequate and ineffective.

And there is no way to avoid Congress. Efforts to clean up present and to prevent future pollution will require federal legislation. The problems demand the kind of money only the federal government can provide. Even quality standards, if they are to be effective, must be national rather than state-wide in application, since varying state standards will simply allow industry to move if the costs of polluting become too high. Indeed, some states would probably keep quality standards low for the sole purpose of attracting industry, even polluting industry, just as some states (notably New Jersey) now attempt to attract industry through low tax rates.

The Committee Jungle

Some aspects of the Congressional system seem designed to frustrate efforts to make real inroads into environmental problems. For example, no single committee in either house handles all legislation relating to ecology—the responsibility is divided among a tangle of competing committees and subcommittees in both houses. In the Senate, committees considering some aspects of pollution legislation include Public Works (particularly the subcommittee on Air and Water Pollution), Commerce (the subcommittee on Energy, Natural Resources, and the Environment and the subcommittee on the Merchant Marine and Fisheries), Interior and Insular Affairs, Labor and Public Welfare, Government Operations, Finance, Agriculture, and a corresponding variety of Appropriations subcommittees. In the House, the following committees claim a piece of the pie: Interstate and Foreign Commerce, Public Works, Government Operations (Natural Resources and Power subcommittee), Merchant Marine and Fisheries, Science and Astronautics, Interior and Insular Affairs, and Agriculture.

The most important consequence of this government by division is not the oft-cited ill effects on policy coordination, for that assumes there is policy to coordinate. The real dire consequence is that few, if any, Congressmen join these committees solely because of the legislative interest in redeeming the environment. Congressmen usually seek appointment to committees which can help them score points with major interests in their home states. Not infrequently these major interests are antithetical to environmental interests.

Air pollution, for example, is the province of the Public Works Committee in the Senate, chaired by Senator Jennings Randolph of West Virginia. Randolph's major interest is not pollution, but protection of West Virginia's coal industry; and he originally joined Public Works because that is the committee which exercises the greatest impact on the mine operators. Not surprisingly, Randolph can be counted on to oppose effective efforts to come to grips with air pollution if the coal industry's interests are adversely affected. The chairman of the Interstate and Foreign Commerce Committee, which handles air pollution in the House, is also a West Virginian with similar interests, Harley Staggers.

Pesticide legislation meets similar obstacles in its prescribed course through Agriculture committees. These bodies are filled with Congressmen from Southern cotton states which are heavy users of DDT. (Two-thirds of the DDT used for farming finds its way to cotton crops.) Senator Everett Jordan of North Carolina is chairman of the Senate Agriculture Committee's Research and General Legislation subcommittee, which handles pesticide bills. Twelve pesticide bills were introduced in Congress in 1969, but none received so much as a committee hearing. Should legislation handled by these committees somehow pass, it would still have to survive the sharp surveillance of the Agriculture subcommittee of the Appropriations Committee in order to be funded, and this subcommittee in the House is chaired by Representative Jamie Whitten of Mississippi, a cotton-state dixiecrat.

Seeking the High Profile

Much of the apparent Congressional activity on the environment does not even concern legislative content but involves a jurisdictional scramble to determine who gets to speak forcefully from a committee chair about clean air and fresh water. The politics of this struggle are byzantine, even by Congressional standards. Senator Henry Jackson, chairman of the Interior Committee, has long concerned himself with resource conservation in the more traditional sense. Recently, however, Jackson has pushed to concentrate water pollution administration within the Interior Department. Senator Edmund Muskie, the leading pollution fighter in the Senate, has in turn fought to keep control of both air and water pollution legislation in his Public Works subcommittee and has favored HEW as the logical focus of federal pollution control.

In 1969, Jackson introduced the National Environmental Quality Act which ultimately established a Council of Environmental Quality in the White House to conduct ecological research and to review and coordinate environmental action by federal agencies. Jackson's proposal was also introduced in the House by Michigan Representative John Dingell, chairman of the Fisheries and Wildlife Conservation subcommittee of the Merchant Marine and Fisheries Committee. Despite the efforts of Wayne Aspinall, chairman of the House Interior Committee, Dingell's subcommittee maintained control of the bill. When the bill passed Congress, Jackson's committee in the Senate and Dingell's subcommittee in the House had established jurisdiction over the Council which then appeared to be the future focus of the government's pollution and environment efforts.

But Senator Muskie immediately responded by introducing a bill to establish an independent anti-pollution agency. The proposed agency was designed to carry out the actual operation of most of the federal anti-pollution and environment programs, just coincidentally, of course, also stripping the Interior Department of its primary role in that area. To the surprise of nearly everybody (most certainly to the surprise of the Interior

Department), in late June President Nixon announced a reorganization strikingly similar to Muskie's proposal, calling for the creation of the Environmental Protection Agency within the White House to run nearly all federal anti-pollution programs. By centralizing the politically popular pollution programs under his exclusive control at the White House, the President can not only take direct credit for pollution fighting efforts, but he can also prevent spending excesses that threaten to upset his budget plans. And though this presidential action has partisan consequences and threatens various Congressional interests, it is unlikely to encounter serious House or Senate opposition, for the reorganization has left the basic political rules of the environmental game intact. The boundaries of the gentlemen's agreement have been respected.

Understanding the nature and causes of this environmental restrictive covenant requires a careful look at the obstacles the Congress would have to overcome if protecting the nation's environment were to become more than a paper priority.

Money Is Thicker than Water

Cleaning up past pollution provides the easiest starting place for a government campaign to save the environment. Restoring the quality of the nation's rivers and lakes, for example, is primarily a matter of spending enough money. And since water pollution control appropriations go for public works that funnel federal funds and jobs into large numbers of Congressional districts, such spending ought to be popular in Congress. Water pollution accounts for more than 80 per cent of all federal expenditures on pollution control and prevention.

Why haven't we gone ahead and rescued our stinking rivers and dying lakes? The fact is that cleaning up America's polluted waterways demands far more funds than either the President or Congress is currently willing to provide.

Last year, Congress appropriated $800 million for water pollution control—a sum which, although $200 million less than authorized, was nonetheless three times greater than President Nixon had requested. Yet, according to Senator Muskie, "If we were to catch up on the backlog of untreated municipal wastes alone—this does not cover industrial wastes—we would have to spend $25 billion over a five-year period. If you were to add industrial wastes, I suspect we would have to double that figure."

In other words, by Muskie's own reckoning, an effective attack on past water pollution alone—excluding efforts to prevent future pollution—would run $10 billion a year, probably half of which would have to be financed by the federal government. And Muskie's is a conservative estimate compared to some environmentalists'.

Therefore, Congressmen serious about pollution abatement must advocate more than a 500 per cent increase in current federal spending on water alone. A quick look at the federal budget reveals areas of fiscal waste, such as military spending, that might be tapped to provide the extra $4 billion annually. But, to date, efforts to substantially reorder prevailing budgetary spending patterns have failed, and there is little hope of success in the foreseeable future.

There are several other possible alternatives for finding funds. Congress could raise federal taxes. But as Senator Tydings explained, "Anyone who believes raising taxes is a viable political alternative at this time is obviously not up for election." Having witnessed the electoral fate of those governors who dared to raise state taxes in recent years, this course is about as popular on Capitol Hill as appearing at peace rallies with Jerry Rubin. No one even talks about it. The feeling is that, while Americans want pollution controlled, they will not react favorably to Congressmen who raise *their* taxes to control it.

Other environmentalists are counting on obtaining the needed funds from additions to federal revenues produced by general U. S. economic expansion (the growth dividend) and an end to the fighting in Southeast Asia (the peace dividend). However, even if a substantial growth-peace dividend should unexpectedly materialize, it is hardly reasonable to assume that a sizable portion of this surplus would be allocated to pollution control. Other problems, pushed by powerful national organizations, are waiting in the wings to compete for additional federal dollars.

The blunt proclamation of George Wiley, director of the National Welfare Rights Organization, suggests the opposition to large expenditures on pollution abatement which groups that do not share the environmentalists' priorities will provide. "Pollution is a priority issue right after we do away with poverty," Wiley commented recently, "and that means a $5,500 guaranteed income for everybody. After we have accomplished that, we can begin the fight on pollution."

It is not difficult to comprehend why America's minority groups and her poor do not believe pollution has the strong claim on federal resources envisioned by the environmentalists. "We haven't given it a very high place on our list of priorities," explained Clarence Mitchell, the NAACP's influential Congressional lobbyist. "Spending on pollution control comes after we take care of the problems of low income, poor housing, and inadequate medical care. Once we get these accomplished, then we can go into the fringe areas such as cleaning up rivers and air."

The predominantly white, middle-class National Education Association, on the other hand, claims to regard the environment as a priority matter. As Stanley McFarland of the NEA put it, "Pollution is high on our list of priorities. We would hope here could be a re-emphasis towards domestic priorities and areas like education as well as pollution." But when asked whether the NEA would favor pollution over education if only a small fiscal dividend materialized, McFarland replied without hesitation that "our major interest, of course, is education. Hopefully, it won't be an either/or question, and we will have enough funds available to do better in a number of areas."

Hard allocation choices, if not either/or questions, are unavoidable. And at this juncture, few of the powerful interests competing for limited federal funds put the environment at the top of their lists. It remains everyone's second choice, and no serious group ever gets past its first choice.

Odds against the environmentalists winning a substantial share of future federal surplus funds are lengthened further by the absence of a stong environmental lobby in Washington. Unlike the traditional conservation lobby with its sportsman-wildlife focus, the new ecologically oriented groups (Friends of the Earth, Environment!, Zero Population Growth, Environmental Action) lack the resources and sophistication to be politically effective. As one House staffer described it, "Once you separate out the old-line conservationists, you don't have much left but long hair and a lot of enthusiasm."

Higher Hurdles

The second stage of any comprehensive strategy for environmental protection—preventing future deterioration—presents even more difficult problems. For, in addition to enormous funding constraints, the movement to save Spaceship Earth from future disaster confronts political roadblocks constructed out of strongly held societal mores, rigid consumer habits, and giant concentrations of economic power.

Public policies to restore the environment will clash inevitably with the nation's demand for unfettered economic growth. Many ecologists contend that efforts to preserve the environment are doomed until economic growth *per se* is prohibited. Thus, John Fisher declared in *Harper's*:

> . . . our prime national goal, I am now convinced, should be to reach Zero Growth Rate as soon as possible. Zero growth in people, in GNP, and in our consumption of everything. That is our only hope of attaining a stable ecology: that is, of halting the deterioration of the environment on which our lives depend.

Paul Ehrlich, going Fisher one better, advocates economic contraction, or "dedevelopment" as he calls it.

These radical dedevelopers often obscure the difference between nonpolluting growth in services, which also adds to the GNP but not to smog, and the forms of basic industrial growth which are inevitable pollutants.

But the fact remains that sound environmental policies and certain kinds of economic growth are incompatible. Industrial production is the source of most environmental harm, and the conflict with powerful economic interests implicit in such a situation constitutes perhaps the most formidable political barrier to meaningful environmental legislation.

Deflating the Air Act

The Air Quality Act of 1967 provides a good case in point. Air pollution is basically a fuels problem. Eliminating dangerous emissions from fuel combustion requires technology to remove the pollutants from the air, or if developing such technology appears improbable, the substitution of safer fuels for harmful ones.

There is general agreement among air pollution control experts that the only way to achieve this elimination is through enactment of federal emissions standards applicable to all major industries, whatever their location. However, any attempt at regulating fuel use strikes at the very heart of America's energy-based economy and quickly arouses the opposition of what Ralph Nader has labeled the "energy establishment." Consisting of the oil and coal industries, the natural gas suppliers, the utilities, the automobile corporations, the atomic energy industry, and major fuel consumers such as the steel industry and railways, this mammoth combine is a virtually irresistible political force.

The Air Quality legislation President Johnson sent Congress in January of 1967 originally called for national industrial emission standards covering every important industry. "Had we enacted and enforced those White House provisions," a Senate staff man working in the area explained, "a huge amount of the air pollution that is poisoning all of us today would have been eliminated."

But according to *Vanishing Air*, the report of Nader's task force on air pollution, Congress never even seriously considered those national emissions standards. When the Air Quality Act of 1967 finally was signed into law by President Johnson, nothing resembling emissions standards remained. The energy establishment didn't have to twist any arms. "There was no need for that," task force director John Esposito declared. "The gentlemen's agreement was in effect; Congress realized it had nothing to gain by taking on this concentration of economic power. They left it to Muskie to disguise this retreat with the trappings of victory."

The only overt industrial lobbying related to the act was undertaken by the coal industry, and the ease and completeness with which the mine owners secured their objectives is instructive. In 1967 when the coal industry learned that HEW was about to publish a report on sulphur oxides, a dangerous pollutant produced by coal combustion, which would justify an executive order requiring air pollution controls at federal facilities by 1968, the industry decided to intervene. The considerable power of the coal industry, under the virtuoso direction of coal lobbyist Joe Moody, was quickly mobilized to discredit the report's findings and delay its publication.

Senator Jennings Randolph of West Virginia was set to work intimidating HEW. In a series of letters to Department officials, Randolph in effect argued that a determination of the health hazards posed by sulphur oxides should be postponed until technology was developed to control sulphur oxides emissions from smoke stacks.

What is incredible is not Randolph's willingness to employ such obviously specious reasoning, but Congress' readiness to accept it. HEW did formally publish its report in March of 1967. But less than five months later, the Congress obliged Randolph by passing the Air Quality Act with a provision that ". . . any criteria issued prior to enactment [that is, sulphur oxide criteria] . . . shall be reevaluated . . . and, if necessary, modified and reissued."

With a minimum of effort, coal had succeeded in burying the government attack on sulphur oxides deep in the bureaucracy, where it languished out of sight for another two years. According to Nader's task force report, Joe Moody, flushed with victory, claimed that he had personally written the entire Air Quality Act of 1967. Most environmentalists on the Hill were inclined to believe him.

Even if the Air Quality Act had been made tougher by Congress, industry pressure could have been applied to the responsible federal bureaucracies in order to prevent rigorous enforcement. Senator Philip Hart of Michigan recently complained that "there are many instances where standards set by administrative agencies are pitifully weak, and even those are often not enforced."

Until now, Congress has been able to mute conflicts between environmental considerations and the demands of unbridled economic growth largely by ignoring the environment. This conscious sacrifice of environmental needs to the sanctity of American industry's prerogatives has been so easy to accomplish primarily because the public has been kept completely in the dark. However, the growing electric power shortage facing the nation promises to bring future clashes between environment and growth out into at least a dim light.

Dilemmas of Power

In the eyes of most environmentalists, electric power plants are both the substance and the symbol of what must be stopped if the planet is to be rescued. To begin with, electric power plants are among the worst polluters in the country. Generating more electricity means generating more air and water pollution, given the present state of technology.

Electric power is also seen as a symbol of the forces threatening the environment. It is the energy source for the innumerable appliances and gadgets that characterize our mass consumption economy—an economy which ecologists believe is inherently destructive. So for both ecological and theological reasons, the electric power front is likely to be the scene of the first open clash in the government-growth battle.

Here, too, the prospects for an environmental victory appear slim. In the first place, a Congressional vote for the environment in this context means a vote against the entire energy establishment. Recognizing the threat this would pose to the fundamental growth ethic that provides American business with both its purpose and its profits, the utilities would undoubtedly find willing allies throughout American industry.

It is questionable whether the utilities would even need to bother forming these alliances given the degree of political influence they already possess. Recently, a Southern Democratic Senator and a Midwestern Republican colleague were chatting in the Senate cloakroom about the states where money from a very few sources can control politics. According to one, "The utilities literally own the politics of nearly all of the Southern States, many Midwestern states, and the Western states with the exception of California." Since Congressional power in many important environmental areas still resides in the hands of Southern committee chairmen, the utilities probably already have sufficient political muscle where it counts to stop any legislation attempts to limit their growth.

In addition, a vote for the environment would be a vote for electric power rationing and restricted use of numerous consumer goods. Is the average American sufficiently committed to ecological preservation to give up his dishwasher, his second TV, and his bedroom air conditioner? John Heritage, Senator Gaylord Nelson's staff expert on environment and one of the architects of Earth Day, is pessimistic:

> I am just skeptical that Congress will do anything to visibly reduce the voters' standard of living. The power question will be the real test of how far the nation is willing to go to preserve the environment. I hope I'm wrong, but my fear is that the force of consumer demand will simply overwhelm the environmentalists' demands for planned power shortages.

Conflicts between environmental concerns and constituent demands create serious political problems for even the most dedicated environmentalists in Congress. Senator Jackson, a long-time conservationist and now an environmentalist, has been the leading Senate supporter of the SST, the supersonic transport attacked for the environmental damage it will cause. Jackson, who represents Washington, the home of Boeing and a state dominated by the aircraft industry, now suffering substantial unemployment, could scarcely have ignored the obvious political imperative he faced.

Similarly, Senator Muskie, probably the most consistent environmentalist in the Senate (it was his major concern long before it was a popular public cause), has been criticized for his support of the building of an oil refinery in Machiasport, Maine, a relatively unspoiled area which would be endangered by pollution if the refinery were built. However, this unspoiled area itself is characterized by the abject poverty and unemployment of its residents, many of whom would derive jobs and income from the refinery. Muskie, obviously caught in a classic political bind, first supported the refinery, then declared himself neutral after he was attacked by conservation groups.

John Dingell, another long-time environmentalist, was recently caught in much the same situation when New York Representative Leonard Farbstein introduced an amendment to pending air pollution control legislation which would, in effect, have banned the internal combustion engine within the decade. Dingell, despite his unquestioned concern over environmental pollution, opposed the amendment. His Detroit area district includes the Ford River Rouge plant, the largest automobile production plant in the world as well as Local 600, the largest local unit of the United Auto Workers. The dilemmas of Jackson, Muskie, and Dingell are examples of the crunch all Congressmen must face at least a few times each year. A look at recent public opinion polls helps explain why the choice made under duress is so often anti-environmentalist in nature. A Gallup Poll published on May 13 asked citizens to choose from a list the three national problems they would like government to devote most of its attention to in the future. Reducing pollution of air and water was mentioned by 53 per cent of the public, second only to crime, which was selected by 56 per cent. However, only a month later the Gallup organization asked citizens to name the most important problem facing the country. This time those interviewed were not given a list to choose from. In apparent contradiction of the earlier results, pollution did not even rank among the top 10 problems, and in fact was named as top problem by only two per cent of those replying.

The contradiction, however, is more apparent than real. Pollution and environmental concerns, despite their recent vogue, simply do not impress most Americans as a problem of major concern to themselves or their country. Pollution control efforts are viewed as good and desirable, the kind of government activity which no citizen of good will could really oppose—but which also will not engender tremendous disaffection or adverse electoral consequences if not seriously undertaken.

It may be that the environmentalists' worst fears will be borne out—that nothing short of disasters will shock the American people and their elected representatives to recognize that preserving our planet requires more effort and self-sacrifice than one day of homage to Mother Nature each year.

3 The Traditional Wisdom of Economic Growth

We consider the readings in this section to be not only representative of the pro-growth conventional wisdom but also to include some of the best thinking available from that perspective. Readers are encouraged to relate the arguments presented here to those of the previous section. The following introductory comments should be of some help in doing so.

The case for economic growth leads off with a selection from the 1970 report of the President's National Goals Research Staff, *Toward Balanced Growth—Quantity with Quality*. The report maintains that while economic growth has been responsible for environmental deterioration, it need not be. Instead of restricting growth, they call for balanced growth, a concept involving measures to reduce pollution and a somewhat reduced "apparent rate of growth." To what extent such solutions are dependent on the political decision-making processes, and on a limited view of the environmental problem (focusing mainly on pollution), are questions that should be considered while reading this selection. It is perhaps the most sophisticated of the statements presented in support of growth.

The next article is by Hans Landsberg of Resources for the Future, an organization that is funded by the Ford Foundation and carries out a great deal of research on natural resource matters. In "A Disposable Feast" Landsberg denies a necessary connection between economic growth and deterioration of the environment, arguing that "any economic system can be made to respond to environmental considerations," primarily by the creation of positive incentives to do so. In the course of the discussion, however, he offers some convincing evidence that it is increasing per capita income, rather than increasing population, that has been the major cause of environmental problems. Landsberg admits that "GNP tells us nothing about 'quality of life,' " but still claims that "there is good reason to believe that well-being will be greater at $4,000 per capita than at $2,000." The fact that environmental problems exist in the U. S. S. R. also is offered as proof of the ideological blindness of pollution without considering the basic similarities in the technocratic, growth-oriented institutions of both countries.

Economists seem to have a tendency to take a very conservative and limited view of society, particularly when it comes to making policy recommendations. Landsberg, in his case, suggests: "One of the most complicated ecosystems is a modern industrialized society. For that reason, in our attempt to cure our ailing environment, we should be certain the label on the prescribed medicine always bears the admonition 'use well before shaking.' "

"The Economy Doesn't Need More People," by John G. Wells, head of the industrial economics division of the University of Denver Research Institute, is primarily concerned with the economic effects of a stabilized population. Wells suggests that the traditional concern of businessmen that a stable population will cause an economic slowdown is not valid, that in fact it will mean a higher standard of living for all. In other words, we will translate production from products provided to support a growing population to increased products for each member of a stable population. The argument, based on well-established economic thinking, would seem to be correct, but leaves us facing the requirement of higher per capita consumption. In order for the environment to tolerate increased production and consumption, the author counts on advances in technology, hoping that "we can both have our population cake and eat it with a technological frosting." This frosting, according to Harrison Brown, may not be much to our taste.

Harold Barnett and Chandler Morse, in their book *Scarcity and Growth,* present a classic analysis of the economics of resource scarcity. Put most simply, they maintain that continuous technological progress holds the solution to natural resource scarcity. The historically decreasing unit costs of natural resource extraction are offered as proof that natural resource availability has not been effectively limited, even though consumption has increased tremendously. One thing to note is that Barnett and Morse develop their theory around resources rather than the quality of the environment; it is the latter which is becoming most critical. A major form of environmental destruction is the use of lower grade ores requiring strip mining and open pits to permit the use of the huge machines that have held the cost of materials down. How would Ian McHarg respond to this statement: "The social heritage consists far more of knowledge, equipment, institutions, and far less of natural resources than it once did. Resource reservation, by limiting output, and thereby research, education, and investment, might even diminish the value of the social heritage"?

Barnett and Morse are dealing with theory; Raymond Ewell in the next article is dealing with the reality of our resource situation. Vice President for Research of the State University of New York at Buffalo and former official of Shell Chemical Company, Stanford Research Institute, and the National Science Foundation, he makes it quite clear in "U.S. Will Lag

U. S. S. R. in Raw Materials" that continued U. S. economic growth is dependent upon appropriating an increasing share of the world's natural resources, mainly through establishing and maintaining "friendly relations" with resource-rich parts of the world. This is necessary, Ewell believes, in order for the United States to remain ahead of the Soviet Union in the "competition for world leadership." With what might seem to be circular reasoning, the author explains that world leadership is necessary in order to assure the supplies of raw materials needed for continued economic growth. The parallel of this point of view with the discussion of imperialism in "The Political Economy of Environmental Destruction" is evident.

The last selection is by Henry Villard, economics professor at the City College of New York. Titled "Economic Implications for Consumption of 3 Percent Growth," it explores what consumption might be like 100 years in the future if the economy continues to increase at the same rate it has for the last 100 years, resulting in a five-fold gain in per capita income by the year 2065. Even though the average income would be $36,000 a year, the author points out that the average standard of living would be well below someone with a $36,000 income today because of the rising cost of personal services, the impossibility of reengineering cattle to produce only fillet mignon, and a number of "adverse consumption" goods, such as cars causing congested roads and tourists jamming scenic areas. Even so, the author remains firm in his support of economic growth. He expresses a concern for the annoying aspects of environmental changes but not with more fundamental ecological and social difficulties. In a way, the paper would seem to support the thesis that the benefits of economic growth and increased personal income may not be as great as expected, and not worth the ecological risks that will be involved.

The National Goals Research Staff: Toward Balanced Growth—Quantity with Quality

. . . Although the contamination of our surroundings can be decried for moral, aesthetic, and health reasons, the resolution of pollution problems will inevitably involve economic policy. This resolution can be effected by research on the causes and consequences of our imposition on

From the Report of the National Goals Research Staff (Washington, D. C.: U. S. Government Printing Office, 1970).

the environment and by political cooperation between the public and private sectors on the policies research devises. Solutions will take time and may cost much, but they can be brought about within the market system as now basically constituted; properly stimulated, the market itself can be among the most powerful tools in a program to alleviate the physical degradation of our surroundings.

In the past, the air and water have been readily available for any purpose. Our production and consumption activities could be carried out without particular concern for conserving our natural gifts.

But as production and consumption have risen (along with population) more and more impositions have been made on the air and waters. These incursions are the consequence of economic growth and the notion that the environment is available at no cost for whatever use we want, including that of disposal of the wastes from economic activity. Thus, when we impose on our air and waters in ways and amounts that use up these necessities, we levy a real social cost. These resources are no longer free to society.

Because private use of the physical environment is available at no cost, the market system which allocates resources operates imperfectly. The imposition made by the individual firm or consumer when waste is spilled does not constitute a purchase by the firm of a unit of its surroundings; therefore, using up the environment is not counted as a cost of production. The supply-and-demand mechanism of the market does not make its adjustments nor allocate resources with full information about all costs that are in reality being incurred by society. With full cost information added to conventional business costs, the supply-and-demand system would channel resources and consumption activities in magnitudes that approximate the best balance of economic activity and pollution.

To attack pollution requires a balancing of the costs of imposing on the environment with the economic benefits obtained from the associated production and consumption. Total prevention of fouling the environment is not achievable. Therefore, a policy goal for "balanced growth" should be to find and enforce a degree of pollution control at which the costs of more control (i.e., prevention) just equal the benefits (or where the quantity of output is balanced by the improvement of the environment). This degree of control, however, must be continually reappraised in light of technological advances and human needs.

A necessary step in forming an antipollution policy would be to determine the costs of pollution. Some costs of environmental degradation, such as the corrosion of buildings by airborne pollutants, are easy to assess, but others are not easy to trace; the links between pollution and disease and the nonmonetary costs from the loss of aesthetic appeal are far more difficult to calculate. But among the most difficult of all costs to assess are those to the ecosystem of the Nation or the world. Considerable and difficult research is needed to provide necessary cost information; much of it is not available today.

Improving information is only a first step. Policies to effect the desired control are the second, but it is most likely that large programs will have to be started without full data on costs—the urgency of the problem will not permit the luxury of waiting for such information.

Much has already been done to reduce industrial pollution, and there is a variety of means available to accomplish more. Direct regulation of polluters and the enforcement of specific standards, as codified by the Clean Air Act of 1963, and the Federal Water Pollution Control Act (as amended) are the principal elements of present policy. Government might also provide subsidies to firms; for example, by tax incentives to stimulate the purchase of equipment with antipollution features. But subsidies (and the enforcement of regulatory standards) require intervention and monitoring by administrative agencies, and the subsidy policy itself raises issues of tax equity: Taxes must be collected to pay the subsidy and it is not clear that fairness is served by imposing them on all taxpayers, including nonpolluters.

Another approach favored especially by economists would be to put prices on pollution. Those who regulate the health of the environment would impose costs (say by direct charges on discharge of pollutants) for contaminating the environment. In other words, the presently hidden social cost of pollution would be made a real dollar cost and assessed to the offending firms, consumers, and municipalities. These costs would then show up in the supply-and-demand relationship that regulates the market system, and the market would then be able to adjust itself to amounts of production, consumption, and pollution believed socially tolerable. Setting the appropriate "price of pollution" would be no simple task, but within broad limits an imperfectly set price is likely to be preferable to no price at all. This policy would bring about a more perfectly functioning market, one that would have both the information and incentive to reduce the environmental problem.

Putting prices on pollution would raise the cost of polluting; that is, of using our scarce resources of air, water, and land for disposal purposes. In the competitive marketplace, raising the price of pollution would tend to lower environmental degradation, as producers would have an incentive to use manufacturing methods that put fewer burdens on the environment. And higher prices for the products that dirty the air and water would induce households to alter their buying habits, to switch to goods and services less likely to cause pollution when they are produced or consumed. Business would be given an incentive to supply products that could be consumed without creating pollution. And with costs and prices adjusted to show the effects of pollution, the issue of equity and fairness would be resolved.

In some instances government action to increase the capacity of the environment to absorb wastes may be cheaper than reducing the creation of waste. Rivers or streams could be augmented or controlled for this purpose, and specified water courses designated to receive pollutants, while others would be protected.

An extreme solution would be intentionally to lower the rate of conventional economic growth. Such a proposal assumes that growth is the cause of environmental degradation and therefore that the cure is to soften the cause. But degradation of the environment is not a necessary consequence of economic expansion; and improving the environment will itself require new equipment—equipment that will be available only from increased output or from diverting resources from other users. Moreover, there is no guarantee that restricting growth would, by itself, reduce pollution; costs and prices—the "signals" of the market—would also need adjustment. Restricting growth would also run counter to other policy objectives: Slowing the rate of expansion would jeopardize full employment, and would hinder the efforts of minority groups and those in poverty to increase their income. Thus, while it may be true that pollution can be associated with growth, it does not follow that consciously curtailing growth represents sound policy.

The diversion of equipment and manpower into cleaning our physical surroundings could lower the country's apparent rate of growth. Although equipment purchased to purify wastes enters the GNP as investment, with full employment, this diversion would probably lower the investment in machines that are more directly productive of marketable output. In time, and if the change in the mix of investments were substantial, the apparent growth rate of the economy could lessen.

The conventional GNP accounts make no allowance for improvements or deterioration of the environment or of the quality of most private products. Since quality changes or increases in benefits are not measured when the air and water are cleaned, these improvements will be unregistered by usual economic accounting methods, just as in the past environmental deterioration has gone unmeasured. It is therefore possible to conceive of some lowering of the conventional growth rate, but at the same time some increase in real wealth if GNP were adjusted for the quality improvement being newly purchased.

Even if some increases in output are sacrificed to serve the purposes of cleaning the environment, this would hardly represent a new event in American economic life. Quantity has been typically modified by concern for quality, and to suggest that ours is an age where the two are in opposition is to misread history and current events. The sacrifice of income in favor of leisure, and the large portion of current output that is used instead of invested, show that quality (leisure) and current enjoyment (consumption) demonstrate that sheer economic growth for its own sake is not and has not been an absolute concern. Therefore, caring for the wholesomeness of our environment can be considered an extension of America's historical concern for quality. . . .

Hans Landsberg: A Disposable Feast

Like most of the problems confronting modern man, those of environmental congestion and pollution resist swift and simple solutions. There is no single cause lending itself to a single cure. Rather, each problem is a synthesis of several, all springing from multiple causes. Much of the current discussion on the environment, however, reveals little or no understanding of these complexities.

The desire to ferret out causes and to swiftly apply remedies has led to speculation about several isolated factors as single determinants of environmental deterioration. Rapid population growth has been placed at the head of the list—an assignment which is deflecting attention from factors of more immediate bearing that are further down the list or absent from it altogether.

Although population growth is the source of many pressing problems, it is not the major factor in environmental pollution now. It is quite likely that it will aggravate the situation in the long run, but to consider the reduction of fertility as the sole response to present problems is to consider a woefully inadequate solution. Effective national measures for enhancing environmental quality must be based on knowledge about the complex and interacting processes that actually underlie pollution, rather than on the popular assertion that population growth alone is to blame.

These processes—the major determinants of environmental pollution in the United States today—can be stated simply: high per capita consumption based on high per capita income, combined with a sophisticated and powerful technology. Some elements of this combination have been recognized and singled out as villains, but as a formula it is incompletely understood.

Electric power generation—a favorite contemporary villain—illustrates the point. Ninety percent of the growth in power generation in the last thirty years has been caused by higher per capita consumption and

From *Resources,* June 1970, published by Resources for the Future.

only 10 percent by population growth. Were we to consider anything above the 1940 level of electric power generation incompatible with sound environment, we would be unable to tolerate a U. S. population today of more than 20 million, assuming current per capita consumption. Or, taking the present population for granted, we would have to slash per capita consumption by 90 percent.

The pervasive effect of income is surprising. The rise of beef consumption in the past two decades would have been only about 35 percent if based on population growth alone. Instead, it rose 120 percent because per capita use went up 75 percent—and now commercial feedlots are a new environmental problem.

The issue is far more complex regarding technology. To begin with, technology-induced problems form a spectrum, extending from nuisançes, inconveniences, and insults to our aesthetic sensibilities all the way to potential threats to the life-supporting capacity of the earth. Similarly, the remedies range from fairly simple and cheap technological and institutional modifications to exceedingly costly ones involving a wholesale revamping of our way of life.

A given technology takes on "good" or "bad" characteristics according to its time, place and purpose. The internal combustion engine, for example, did not come under indictment as a polluter of air until recently, but preoccupation with the motor vehicle as a safety hazard goes back to its very beginnings. (That this long-term concern has not produced effective safety measures is another matter.)

Only ten or fifteen years ago did the growth in the number of motor vehicles justify their inclusion among the major sources of pollutants. The increase of some 35 million vehicles in the last decade has obviously overtaxed the assimilative capacity of the air, especially over metropolitan areas. This may have happened longer ago. Our increasing ability to measure and evaluate environmental conditions reveals that sometimes the assimilative capacity has been exceeded without our recognizing it.

Certain polluting effects, of course, are by-products of the heat produced by the engine. Thus, the transition from the Model T to the souped-up 300-horsepower high-compression model of the contemporary scene contributed greatly to making motor vehicles a serious problem.

Since air has a vast capacity for harmlessly absorbing emissions of gasses, there is nothing wrong with using it as an assimilator of wastes, provided we recognize the damage threshold and the cost of the cure if we go over this threshold. Demands for zero emissions or zero tolerance—which easily turn into demands for zero motor vehicles—needlessly compound a difficult situation.

It is the internal combustion engine in its present form that creates the difficulty. If we are lucky, modifications of engine and fuel will correct it at a moderate cost and a minor sacrifice in "performance." Failing this, new propulsion systems will be required, and this would entail rather far-reaching changes in the economy. But in either event, the difficulty is well defined and the problem is tractable.

Not all technologies are amenable to such painless modification. Some, if pursued without check could undermine our life supports, and the remedies are less obvious. This leads many to broaden their concern into condemning "the system" under which technology flourishes, including the economic and political organization, and sometimes a vaguely conceived image called "modern man."

In most instances the culprit is an amorphous conglomerate of these factors, since they are difficult to separate. For example, unless the producer who advertises environmentally harmful items is matched by a purchaser who is willing to be persuaded, no untoward consequences will emerge.

Our economic structure is based on a market system using costs, prices, and profits as guides to resource allocation. Few would contend that it is free of defects. But there has been no other system in history—nor is there one on the horizon—that has managed so well, at least cost, to allocate resources among myriads of possible and competing end-uses.

When it comes to disposing of wastes, however, we have no semi-automatic controls analogous to those regulating production and consumption. Indeed, here the system often works in reverse. Striving for least cost for themselves, producer and consumer both tend to dispose of waste in ways that impose the greatest cost on society. In short, the market economy is a reasonably satisfactory organizing principle for allocating resources in production, but it does not help us—and often hinders us—in organizing the handling of wastes at least cost to society.

Until recently this deficiency was of little significance. In earlier times the capacity of the environment to assimilate waste was quite adequate for the then prevailing levels of population, income, and technology. Consequently, the environment could legitimately be treated as a "free good," and limitations on its use were not necessary.

This has been true not only in market economies, but also in societies following totally different economic philosophies. Environmental pollution is a problem in the Soviet Union and in the East European satellites. In these nations too, production and consumption patterns have imposed strains on the environment for which there economic systems provide no corrective. In the Soviet Union, Lake Baikal is the most publicized example; recent banning of DDT is another.

Any organizing principle of production other than one that explicitly assigns a value to environmental factors will tend to use these cost-free aids of production so intensively that eventually symptoms of excessive use appear—namely pollution. Pollution will tend to occur sooner where incomes are high (and per capita production and consumption are also high) and later where incomes are low. Any economic system, however, can be made to respond to environmental considerations, and that is the real challenge.

Economic growth and its yardstick—GNP—also have fallen into disrepute as more of us begin to perceive the connection between high income and consumption and environmental problems. But economic growth need not consist of extras, frills, and planned obsolescence. It can also consist of public goods, including improved environmental quality.

Economic growth should stand for increased options for everybody. In principle, therefore, it is something to embrace. It means moving from spending 70 percent of the household budget on food, as in much of Asia, to spending less than 20 percent, as in this country.

We should take a close look at the consequences of halting growth merely on the ground that wastes are increasing. And it is foolish to attack GNP—a useful indicator of some of the economy's characteristics, if not of others. GNP tells us nothing about "quality of life" or happiness because it was never intended to register values that are not bought and sold within the economy. Even so, there is good reason to believe that well-being will be greater at $4,000 per capita than at $2,000. Those at the lower end of the scale will be hard to convince to the contrary.

Then there is the corporation. Since it lives by the profit motive, it obviously exploits any cost-cutting opportunity, especially free use of the environment. But this opportunism is not unique to the private corporation. The Soviet Steel Trust behaves exactly as U. S. Steel does. In both instances, only the imposition of specific constraints on the producer brings about consideration for the environment.

The imposition of charges high enough to compensate for environmental damage would stimulate a search for a technology that would help the corporation reduce these charges or escape from them totally. But air, water, and land pollution are alternative ways of managing waste disposal; hence the charges must be structured to prevent the air polluter from turning around and becoming a water polluter, or vice versa.

With this qualification, there is no reason to believe that competition cannot become a help rather than an obstacle to environmental enhancement. In the search for new policies we have barely scratched the surface of a large potential. The corporation, after all, has come to terms with industrial safety, with minimum wages, with the end of child labor, and with many other institutions that are not in its short-run interest but that society has imposed on it.

It is a perfectly valid contention that the corporation can be made responsive to policies designed to protect the natural environment. The real difficulty lies in translating concepts into a working system.

Everyone knows the problem created by the nonreturnable container. No amount of exhortation will convince the industry to change to a returnable one, nor the consumer to deposit the empties in ways that facilitate their collection and reuse. In both cases it is a matter of cost. The producer finds it cheaper not to be bothered by collection and reuse. The consumer finds it more convenient—cheaper, that is, in his own way—to dump the container wherever he has emptied its contents.

In this situation, our economic system works in a perverse way because we do not put a price on access to the environment and because different parts of the economy have developed at different rates and are out of whack. The considerable rise in wages and the costs of services have made collecting and transporting wastes uneconomical. At the same time, no cost inhibition keeps people from dumping the containers all over the landscape.

Logically, this calls for modification of the incentives. If a cost is attached to the dumping it will (1) keep consumers from engaging in it, (2) make it worth somebody's effort to collect those containers that are dumped nonetheless, and (3) by encouraging collection, sufficiently reduce the cost of reuse to make it competitive with new material. It would even pay society to subsidize the operation, if that should become necessary to close the cycle.

This approach is applicable to other solid waste problems. Perhaps modification of ownership characteristics will give us something like a "returnable automobile"; returnable not for constant repairs but in the sense that a residual of ownership remains with the producer and that he must accept responsibility for the vehicle when it has become unserviceable. In that event, he would design the parts of a vehicle for maximum reuse. In addition, he would have an incentive to facilitate collection. This probably would raise the cost of automobiles, but anyone who believes that environmental improvement can be had without cost engages in wishful thinking.

The best disposal policy is not to generate anything to be disposed of; to close the production-consumption cycle without a leak to the environment. This is not possible, of course, for materials are like fuels; and recycling itself commonly requires energy and may generate pollutants of its own. The problems posed by disposal of waste heat, by emission of carbon dioxide to the atmosphere with its long-run potential for climatic change, and by disposal of nuclear fission products, are among the most serious environmental issues of the future and will require a vastly increased research effort. But for most solids, reuse will be a realistic goal.

Some forms of environmental injury will require more drastic measures—situations in which the restraint of a price tag is not enough. The use of certain pesticides is a good example. Though our knowledge is incomplete, prudence demands that, with a few exceptions like malaria control, we prohibit the use of DDT rather than rely on a tax to restrain its application.

In most cases, however, intelligent use of economic incentives will do the trick in a manner more compatible with individual freedom, although it will take real political strength to bring about the necessary changes. In this struggle, as in any other, indignation and rational analysis can make a good pair. The first provides momentum; the second protects us from hastily trying to implement unacceptable solutions.

The ecologist insists on the interrelatedness of the components making up an ecosystem. One of the most complicated ecosystems is a modern industrialized society. For that reason, in our attempt to cure our ailing environment, we should be certain the label on the prescribed medicine always bears the admonition "use well before shaking."

John G. Welles: The Economy Doesn't Need More People

Until very recently we have generally accepted population growth as being good for our cities, our states and for the nation as a whole. We have equated population growth with conquering the frontiers of our nation, with more jobs, with business prosperity, and generally with the economic and social well-being of the entire country.

Now, however, doubts are being raised about how good further population growth really is. These doubts are primarily directed at further growth of our larger metropolitan areas. We are being told that there are two sides to such growth: costs as well as benefits.

For example, we are told population growth means higher taxes to pay for needed schools, fire stations, streets and other public services. We are told population growth and concentration in major cities leads to increased crime rates and loss of open space. We see that population growth increases congestion in the cities and in recreation areas.

But the biggest cost being talked about now is pollution: Air, water, land, noise, visual and thermal pollution. The prophets of gloom and doom say the ability of our planet to support human life will be destroyed unless we stop polluting the ecosystem. As an increasing number of prominent scientists join this group, many of us have become more than a little worried. While their case has not yet been proven, what if they are right?

One way to ease pollution and other costs of growth is to stop population expansion. But like any other suggestion for change, the idea of a stable population in the U. S. is not very popular. It tends to scare those who believe that either "we grow or we die."

Zero Population Growth

Nevertheless, an increasing number of prominent Americans, including Secretary Robert Finch of the Department of Health, Education and Welfare, and Lee DuBridge, the President's Science Adviser, are advocating a stable population for the U. S.—what is known as "zero population growth."

Therefore, it seems timely to look at what the consequences might be of zero population growth for the economy of the United States. Much of what follows is based on a study made late last year by Dr. Stephan Enke, an economist at Tempo, a division of General Electric Co.

At present, each family in the U. S. has an average of about three children. To achieve a stable population given certain assumptions including lengthening of life spans, the average number of children in each family would have to be reduced to approximately two. This means a reduction of approximately one-third in the number of children born to the average family.

The startling thing, however, is that if beginning in 1975 each new family limited itself to two children (an unlikely event), the Tempo Study estimates it would take until about the year 2050 for the U. S. population to level off. Analysis indicated the population would grow from the present 205 million to about 293 million before it stabilized. In other words, it would take about 75 years for the population growth to gradually slow down and to stop once we had achieved the two-child family. The time lag is caused by the greater number of young people than older people in our population.

If we look at the year 2060, the impact of different birth rates is dramatically illustrated. If a two-child family were achieved in 1975, we would have in 2060 three people for every two now. In contrast, if present birth rates were maintained, there would be seven people in 2060 for every two now.

As to what the consequences might be for the economy, relatively few economists have examined this question. While no one can predict the future with accuracy, those economists who have examined the question generally agree that a zero population growth would mean a higher, not a lower, standard of living for the average American. This would also mean a growing economy—in spite of zero population growth—because each individual would have increasing buying power.

The Tempo Study arrived at this conclusion. It estimated that living standards in 2060 would be about 20% higher with achievement of a two-child family in 1975 than if population were to grow at current rates.

Reasons for Higher Standard of Living

The major reasons a zero population growth would be expected to result in a higher standard of living are twofold. First, the average family, having only two children rather than three, would tend to save more money. According to well accepted economic theory, greater savings would result in more capital invested for each worker, producing higher labor productivity. And this would help produce the higher standard of living.

Second, the two-child family would result in a significantly lower proportion of the population falling in the young age group. There would be more income per person because there would be proportionately fewer people dependent on those who worked.

While the average American is estimated to be better off with zero population growth, it should be noted that the Tempo Study calculates total gross national product (all goods and services produced) would be roughly twice as large in 2060 if population continues to expand at current rates as opposed to the leveling off of population starting in 1975. This and other ramifications of a stable population have received little attention.

Economists have generally felt that the larger the market served, the cheaper the cost of production. Would not a growing population then result in more cost savings than a stable population? Perhaps in certain instances. However, with over 200 million people already in the U. S., this market is believed to be large enough to support most cost savings resulting from large scale economic activities. In fact, it is quite possible that if the market gets larger, there will be what economists call "diseconomies of scale." For example, it might be cheaper to supply New York City with food if the city had one-half its present population. The lesser congestion might speed delivery of foodstuffs sufficiently to decrease the cost of food.

In addition, the larger the population, the greater the amount of land that has to be placed into agricultural production to feed the population. This means that with each added increment of population, poorer land will be used for production of crops. This would tend to increase the cost of food stuffs. The same reasoning would apply to other land-based resources. And, at least theoretically, the smaller the family the better educated would be the children.

Real estate is one sector of the economy that would not prosper as much in a stable population. The faster the population growth in a prosperous country, generally the faster land values tend to rise. If our population were to level off, real estate speculators would stand to gain less capital appreciation, although there would still be changes in real estate values due to shifts in location of people and businesses.

It is important to note, however, that rises in real estate values do not increase the wealth of a nation. They merely result in transfers of wealth from one segment of a society to another. And as it would take about 75 years for the population to level off if a two-child family were established as the norm tomorrow, it is doubtful that many present real estate speculators would be still in business.

In summary, when we balance the positive and negative factors, it is apparent that the average American, as seen by the Tempo Study, would be better off with a stable population than with a fast growing population.

Poor People Pollute Less

Now let us consider technology under a stable population. The amount of technology used in a society generally is a key factor in determining the standard of living or gross national product of that nation. Unfortunately, however, the amount of pollution occurring in a nation has generally been a function of the amount of technology used. The richer the nation, the greater the pollution.

Hopefully, anti-pollution efforts will be successful in allowing a nation to be rich without excessive pollution. However, sooner or later it seems that something has to give. Either we stop population growth or we reduce our standard of living. Poor people pollute less, so the world could support more of them from an ecological standpoint.

Some ecologists say Earth could not tolerate a raising of the standard of living of the world's population to that of the U. S. They maintain that the planet's ecosystem, especially the oxygen producing subsystem, simply could not stand the degree of pollution that would result. The situation in the year 2000 would be even more alarming to the ecologists since by then the world's population and the U. S. standard of living each are expected to have doubled.

Ironically, all of the nations in the world aspire to a higher standard of living. According to one estimate, the world's standard of living would have to be increased seven times today, and approximately twenty-eight times by the year 2000, in order to equal that of the U. S. It is easy to understand why such growth prospects terrify the ecologists.

If we can learn how to apply large amounts of technology without destroying the ability of our planet to support human life, then perhaps we can both have our population cake and eat it with a technology frosting. It is too early now to tell how this might work out. Meanwhile, the time has come to take a dispassionate look at those consequences of continued population growth in the U. S.

U. S. Damaging the Planet

Perhaps it is also time to take a fresh look at the meaning of human life. It appears from the foregoing that we may have to make decisions between numbers of people and standards of living. Other difficult questions are also being asked. For example, it has been advocated by certain Americans that we extend foreign aid only to those developing nations willing to control their population. However, it has been estimated that one American will pollute as much as 30 to 300 persons in India. From the ecologist's point of view, it would seem that population control in the U. S is equally if not more important to the preservation of our planet than population control in developing nations.

Paul Ehrlich has said in *The Population Bomb* that it is possible (but not likely in his opinion) the dangers of overpopulation have been overestimated. If this turns out to be the case after a stable population has been achieved, there is no problem if man decides he wishes to resume his multiplication.

While few things are certain in forecasting, economic thinking to date indicates that stable population in the U. S. would likely result in a higher standard of living for the average American and in continued economic growth. For those businesses that might be adversely affected, there would be approximately a 75-year adjustment period. Of much greater importance, a stable population would significantly ease future environmental pollution.

Harold J. Barnett and Chandler Morse: Scarcity and Growth— The Economics of Natural Resource Availability

---

Man's relationship to the natural environment, and nature's influence upon the course and quality of human life, are among the oldest topics of speculation of which we are aware. Myth, folktale, and fable; custom, institution, and law; philosophy, science, and technology—all, as far back as records extend, attest to an abiding interest in these concerns.

From *Scarcity and Growth—The Economics of Natural Resource Availability* by Harold J. Barnett and Chandler Morse. Published by the John Hopkins Press for Resources for the Future, Inc. Reprinted by permission of the publisher.

The Doctrine of Increasing Natural Resource Scarcity

The past two centuries—the period of industrial revolution, emergence of science, and population explosion—have witnessed a great broadening and deepening of interest in natural resources. An influential expression of this growing interest was that of British classical economics, early in the nineteenth century, with its doctrine that an inherently limited availability of natural resources sets an upper bound to economic growth and welfare. Later, there was the Conservation Movement in the United States, which took shape around the turn of the present century. Arising out of concern over natural resource scarcity and a consequent endeavor to formulate policies for the use of the extensive public domain, this movement provided broad, vigorous, and influential expression and political leadership. The interests, but not the vitality, of the Conservation Movement, survive in a vast current literature of scientists, engineers, social analysts, educators, journalists, businessmen, public officials, and adherents of a wide variety of academic disciplines. The occurrence and economic consequences of natural resource scarcity, and their social and policy implications, run like strong threads through the variegated fabric of contemporary public concern over natural resources. The doctrine of increasing scarcity and its effects has achieved remarkable viability.

The classical economists—particularly Malthus, Ricardo, and Mill—predicted that scarcity of natural resources would lead to eventually diminishing social returns to economic effort, with retardation and eventual cessation of economic growth. Indeed, classical economic theory acquired its essential character, and for economics its reputation as the "dismal science," from this basic premise. In a somewhat different formulation, the scarcity idea also entered the theory of natural selection when Darwin, acknowledging a debt to Malthus, saw competition for limited means of survival as the determinant of biological evolution.

The Conservation Movement accepted the scarcity premise as valid for an unregulated private enterprise society. But, rejecting laissez faire, at least so far as activities connected with natural resources were concerned, they believed that the trend of social welfare over time could be influenced by the extent to which men conserved and managed resources with an eye to the welfare of future generations. The leaders of the Conservation Movement proposed that society, taking thought for the consequences of its actions, should forestall the effects of increasing scarcity by employing criteria of physical and administrative—not economic—efficiency. They argued that government intervention could improve on the untrammeled processes of private decision making with respect to natural resources, and that public policies should be devised with this end in view. Their willingness to employ public power as a check on business freedom, signifying as it did a certain disillusionment with the principles of laissez faire, made the Conservation Movement something of a catch-all for interventionist ideas of all kinds. This helps to explain its many-sided character. But the core of the Movement was concern for the effect of natural resources, and especially natural resource policy and administration, on the trend of social welfare in a world subject to increasing resource scarcity.

Our principal concern is with economic doctrines of increasing natural resource scarcity and diminishing returns, and their relevance in the modern world. This reflects a professional bias, but also a conviction that serious consideration of the social, and more qualitative, aspects of the natural resource problem must be secondary—that is, must follow understanding of the more quantitative economic aspects. For, if growth and welfare are inescapably subject to an economic law of diminishing returns, the necessary social policies and the moral and human implications are surely different than if they are not. Alternatively, if there is reason to believe that man's ingenuity and wisdom offer opportunities to avoid natural resource scarcity and its effects, then the means for such escapes and their moral and human implications become the center of attention.

The problem we treat, it will be apparent, lies in the realm of historical growth economics, not of static efficiency economics. The latter enters our analysis on occasion, but only in a subsidiary fashion. Our framework is that of the classical economists in their theorizing about the trend of output per capita over the long term. Our main effort, therefore, is directed to a thorough examination of the conceptual and empirical foundations of the doctrine of increasing natural resource scarcity and its effects. . . .

Resources in a Progressive World

[When the assumption of social and technological constancy is removed] it then becomes virtually impossible to postulate a realistic set of conditions that would yield either generally increasing natural resource scarcity or diminishing returns in the social production process as a whole.

Recognition of the possibility of technological progress clearly cuts the ground from under the concept of Malthusian scarcity [an absolute limit to natural resources]. Resources can only be defined in terms of known technology. Half a century ago the air was for breathing and burning; now it is also a natural resource of the chemical industry. Two decades ago Vermont granite was only building and tombstone material; now it is a potential fuel, each ton of which has a usable energy content (uranium) equal to 150 tons of coal. The notion of an absolute limit to natural resource availability is untenable when the definition of resources changes drastically and unpredictably over time.

What technological progress does to the Ricardian scarcity hypothesis [unlimited resources, but not homogeneous] is less clear. We take account of two possibilities. One—which we call the strong hypothesis—is that the economic quality of resources undergoes decline despite the occurrence of technological progress. We cannot, unfortunately, test this hypothesis directly by measuring changes in the economic quality of resources—that is, in natural resource scarcity. However, we can measure changes in the cost of extractive output, an imperfect but reasonably acceptable stand-in for natural resources themselves. When this is done for the United States for the period 1870–1957, the indexes (1929 = 100) of labor-capital input per unit of extractive output are as [shown in Table 1].

The evidence mainly shows increasing, not diminishing, returns. The

Table 1.

	Total Extractive	*Agriculture*	*Minerals*	*Forestry*
1870–1900	134	132	210	59
1919	122	114	164	106
1957	60	61	47	90

cost per unit of total extractive product fell by half during the period. Of agriculture and minerals—the two major components, which account for 90 per cent of the total—the latter fell even more. Only forestry gives evidence of diminishing returns; cost per unit of product rose from the Civil War to World War I. But since World War I, when an aggravation of scarcity due to further economic growth might have been expected, diminishing returns have given way to approximately constant (or slightly increasing) returns.

On the whole, our strong hypothesis of natural resource scarcity fails. True, the increase in the absolute cost of forest products can be taken as evidence tending to confirm the validity of the Ricardian hypothesis in this particular sector. During the period up to 1919, the exhaustion of the better and more accessible stands of trees led to increasing costs. In the subsequent three decades, however, the rise in costs had begun to produce three effects: an introduction of cost-reducing innovations, conversion of wood wastes into usable products, and a shift to wood substitutes. The first and second of these contributed primarily to the observed stabilization in unit costs; the third helped to moderate the effect of the previous rise in costs on the rest of the economy.

Our second—and weaker—formulation of the Ricardian hypothesis is that, in the extractive sector, the decline in resource quality will partly nullify the effect of economy-wide technological advance; and that costs of a unit of extractive goods will, therefore, rise relative to unit costs of nonextractive goods. The evidence—again for the United States, 1870–

1957, and using indexes (1929 = 100) of labor-capital input per unit of output—is as [shown in Table 2].

Increase in natural resource scarcity should, according to the weak hypothesis, prevent the cost of extractive output from falling as much as that of nonextractive output. But, again with the exception of forestry, this did not occur. Relative unit costs of total extractive goods, and of agricultural goods, have been constant, and those of minerals have fallen; those of forestry alone have risen.

Developments like these are to be anticipated. We may expect that particular extractive products will undergo cost increases from time to time. But we may also expect that substitutes will make their appearance (as in the case of fuel and lighting), and that cost-reducing innovations will begin to put in an appearance (as in production of food grains). Increasing costs for particular extractive products, therefore, do not signify increasing costs for extractive output as a whole, let alone for the aggregate of all goods and services. Either costs are brought down, or the product market is lost as costs rise and substitutes appear, or both developments occur.

Table 2.

	Total Extractive Goods Relative to Nonextractive Goods	*Agricultural Goods Relative to Nonextractive Goods*	*Minerals Relative to Nonextractive Goods*	*Forest Products Relative to Nonextractive Goods*
1870–1900	99	97	154	37
1919	103	97	139	84
1957	87	89	68	130

These developments, it is important to realize, are not essentially fortuitous. At one time they were, but important changes have occurred in man's knowledge of the physical universe over the past two centuries, changes which have built technological advance into the social processes of the modern world. A major proportion of final output in Malthus' time represented mechanically transformed products derived initially from a heavily agricultural, and still somewhat primitive, extractive sector. Today, chemical processes and molecular transformation—not to say atomic—have greatly broadened the resource base. Such ubiquitous materials as sea water, clays, rocks, sands, and air have already become economic resources to some degree, and constitute major plateaus of virtually constant physical properties and—under the prodding of continual research and development—increasing economic quality.

The scientific age differs in kind, and not only in degree, from the preceding mechanical age. Not only ingenuity but, increasingly, understanding; not luck but systematic investigation, are turning the tables on nature, making her subservient to man. And the signals that channel research effort, now in one direction, now in another—that determine innovational priorities—are usually the problems calling loudest to be solved. Sometimes the signals are political and social. More often, in a private enterprise society, they are market forces. Changes in relative costs, shifts of demand, the wish to develop broader markets—all aspects of growth—create problems which then generate solutions. Technological progress is no longer a mere by-product of man's efforts to win a decent living; it is an inseparable, organic component of that process.

Thus, the increasing scarcity of particular resources fosters discovery or development of alternative resources, not only equal in economic quality but often superior to those replaced. Few components of the earth's crust, including farm land, are so specific as to defy economic replacement, or so resistant to technological advance as to be incapable of eventually yielding extractive products at constant or declining cost. When coal, petroleum, hydroelectric power, and the atomic nucleus replace wood, peat, and dung as sources of energy; when aluminum yields its secrets to technology and is made to exist, as never before, in the form of metal; when the iron in taconite, once held there inseparably, becomes competitive with that in traditional ores—when all this happens, can we say that we have been forced to shift from resources of higher to those of lower economic quality?

We think not; the contrary is true. And we doubt that it is proper, in long-term, empirical growth analysis, to ask what would have happened to the economic quality of natural resources in the absence of technological progress. For the technological progress that has occurred was a necessary condition for the growth that has occurred, and if the former is ruled out the latter cannot appropriately be taken as a given fact. The strength of the demand-pull on the resource base, therefore, and the resources that would have been used in the absence of progress, are not meaningfully determinate. Hence, we are unable to make a quantitative statement about what would have happened to the cost of extractive output in the absence of sociotechnical change, beyond the analysis of our highly constrained classical models. Besides, our curiosity about a stagnant world is not very pressing, given the rapid change of the one we live in. Under the circumstances, it is more interesting and useful to reformulate our views concerning the influence of natural resources so that they will fit the changing world.

Welfare in a Progressive World

A reformulated hypothesis for the modern world must recognize that the exceedingly varied natural resource environment imposes a multitude of constraints—social no less than physical—upon the processes of economic growth. This presents expansionist man with a never-ending stream of ever-changing problems. But modern solutions to particular problems seldom entail increasing costs. Advances in fundamental science have made it possible to take advantage of the uniformity of energy matter—a uniformity that makes it feasible, without preassignable limit, to escape the quantitative constraints imposed by the character of the earth's crust. A limit may exist, but it can be neither defined nor specified in economic terms. Flexibility, not rigidity, characterizes the relationship of modern man to the physical universe in which he lives. Nature imposes particular scarcities, not an inescapable general scarcity. Man is therefore able, and free, to choose among an indefinitely large number of alternatives. There is no reason to believe that these alternatives will eventually reduce to one that entails increasing cost—that it must sometime prove impossible to escape diminishing quantitative returns. Science, by making the resource base more homogeneous, erases the restrictions once thought to reside in the lack of homogeneity. In a neo-Ricardian world, it seems, the particular resources with which one starts increasingly become a matter of indifference. The reservation of particular resources for later use, therefore, may contribute little to the welfare of future generations. The social heritage consists far more of knowledge, equipment, institutions, and far less of natural resources, than it once did. Resource reservation, by limiting output, and thereby research, education, and investment, might even diminish the value of the social heritage.

Population growth constitutes a special problem. Living space on, or effectively near, the earth's surface is limited. But if living space is the ultimate limiting factor, the notion of Malthusian scarcity is no longer what it was a century and a half ago. The space limitation seems more likely to become manifest in crowded living conditions, a changed environment, an altered quality of life, than as increasing unit costs. For this reason, man may eventually undertake to limit his numbers, not by the operation of positive Malthusian checks but voluntarily, to avoid the qualitative effects of overcrowding—or, more immediately, in the less developed nations, to improve their prospects of increasing capital per head and the rate of growth of output relative to population. Malthusian scarcity would thus be transformed from a problem of subsistence, the lower limit of man's survival, to one concerned with the quality of life, with raising the upper limit to man's total welfare.

And this, in effect, is also true of Ricardian scarcity. For if the increasing scarcity of particular resources generates quantitatively adequate antidotes to the increasing scarcity of resources in general, the cost aspect of potential diminishing returns ceases to be of dominant interest. With many alternatives from which to choose, the criteria of choice become a central concern, for it is these criteria, and the effectiveness of man's decision-making procedures, that will mainly affect the shape of total welfare over time.

This is a problem that, in an intuitive way, concerns contemporary conservationists. To an increasing extent, as they partly realize, the problems of natural resources are qualitative. The difficult questions now are not whether physical and economic problems can be solved, but which problems to solve and how to solve them. The kinds and qualities of changes in the environment, in the social production process, and in commodities; the composition and the allocation of benefits and costs; the standards and the procedures by which alternatives are to be evaluated—these, rather than the cost of an increment to the pre-existing product mix, have become of increasing concern. As man extends his mastery over output and its cost, it is inevitable that these social problems will acquire increased significance and receive greater attention. Whether the attention they receive will suffice to assure continuing improvement in the quality of life is now the open question.

Raymond Ewell: U. S. Will Lag U. S. S. R. in Raw Materials

The high level of affluence in the U. S. is consuming industrial raw materials at a very high rate—and at a steadily increasing rate as our level of affluence continues to rise. Most of these industrial raw materials are of mineral origin, such as petroleum, natural gas, iron ore, copper, chromium nickel, aluminum, and the like. In addition, we get lumber, wood pulp, natural rubber, and some vegetable oils from forest resources, and cotton, wool, and a few other industrial raw materials from agricultural sources. The only important industrial raw materials we get from the ocean are salt, magnesium, and bromine (other than offshore petroleum production). In this discussion I am referring only to nonfood raw materials.

Our domestic mineral and forest resources are being rapidly depleted—to the point where the U. S. is being forced to import more and more of her industrial raw materials. This trend has been going on for many years, but at an accelerating rate during the past 20 years. I estimate that out of the 36 most important industrial raw materials consumed by our manufac-

---

turing industries, the U. S. is self-sufficient in only 10 of these materials, and must import all or part of requirements of 26 materials.

Large-scale importation of so many materials vital to our national well-being leads to many economic and political problems and obviously has an adverse effect on our international balance of payments. As our shortages of industrial raw materials develop still further, this may have the effect of slowing the rate of increase of our standard of living and eventually lead to a leveling off and still later to an actual decline in our standard of living. This may happen in the next 20 to 30 years. The U. S. is not exactly scraping the bottom of the barrel yet, but the remaining known reserves of raw materials per capita are much less than they were 20 years ago and very much less than 50 years ago.

This prospect is magnified even more by the expected growth of U. S. population. The population of the U. S. is now 205 million and currently growing at 1% a year. If this rate continues, our population with be 275 million in the year 2000. It will obviously be more difficult to provide a high standard of living for 275 million than for 205 million persons.

If it were not for the competition for world leadership between the U. S. and the Soviet Union, the most serious effect of this trend of affairs on the people of the U. S. would be a gradual decline in our material standard of living. That would not be too disastrous. We could undergo a considerable decline in our material standard of living and still survive.

Competitive Position

The competition, nonetheless, for world leadership between the U. S. and the Soviet Union puts an entirely different light on this picture. While the raw material position of the U. S. is declining, the raw material position of the Soviet Union is becoming stronger. Of the 36 major industrial raw materials, the Soviet Union is self-sufficient in 29 and needs to import only seven of these materials.

Today the Soviet Union has much greater reserves of mineral and forest resources than the U. S., and known reserves of mineral resources in the Soviet Union are steadily increasing as new deposits are discovered. As recently as 1950 the known reserves of mineral resources in the U. S. exceeded those of the Soviet Union; during the past 20 years the mineral resources of the U. S. have gone down while those of the Soviet Union have gone up. I would estimate that the cross-over point occurred about 1955. The Soviet Union, for example, now greatly exceeds the U. S. in such basic mineral resources as petroleum, natural gas, iron ore, aluminum ore, lead/zinc ore, chromium, and manganese, in addition to other vital materials such as gold, platinum, and diamonds. In addition, the Soviet Union has incomparably greater forest resources than the U. S. or even the U. S. and Canada combined. . . .

Industrial Production

Even though the Soviet Union in 1970 has much greater reserves of industrial raw materials than the U. S., our total industrial production greatly exceeds that of the Soviet Union. Although the ready availability of raw materials at economic prices is one of the principal determinants of industrial production, there are other important determinants, too. Our deficiency in raw materials vis-a-vis the Soviet Union in 1970 is more than compensated by the following factors:

1. Superior technology in nearly every industry.
2. Better research and development—the source of our superior technology.
3. More productive labor.
4. More efficient organization and management in industry.
5. Proximity of Canada and Mexico as ready sources of industrial raw materials.

The importance of Canada (and to a lesser extent, Mexico) cannot be overemphasized. Canada is a vast storehouse of industrial raw materials that are virtually as accessible as if they were within our own borders. Without Canada our raw material position would be much more precarious than it is now. Moreover, Canada is still in an exploratory phase, geologically speaking, whereas the U. S. (except Alaska) has been quite thoroughly explored.

Although the above five factors are presently keeping us ahead of the Soviets in industrial production, this superiority will become increasingly difficult to maintain as the raw material gap between the U. S. and the Soviet Union continues to widen in the years ahead. By 1990 or 2000, the Soviet superiority in raw materials will become so overwhelming that we may not be able to maintain our lead in industrial production, including both civilian and military production, unless we now ensure adequate future supplies of raw materials.

Another aspect of the industrial competition between the U. S. and the Soviet Union is our differing emphasis on types of goods. Although total industrial production in the U. S. is greater than in the Soviet Union, a large part of U. S. industrial production is comprised of luxury consumer products, which the Soviets do not even make or produce in only token quantities. Automobiles for everybody, the vast array of electrical appliances, motor boats, snowmobiles, swimming pools, expensive sports equipment, instant foods, cosmetic *ad absurdum,* elaborately equipped single-family homes, *et al.* are examples of luxury consumer products that make up the "superaffluent society." Such products comprise a large part of U. S. industrial production but are made on a very small scale or not at all in the

Table 3.
The Soviet Union Is Self-Sufficient in 29 Major Industrial Raw Materials*

United States		Soviet Union	
Self-sufficient	Not self-sufficient	Self-sufficient	Not self-sufficient
Coal	Iron ore	Coal	Vanadium
Uranium	Copper	Uranium	Tin
Molybdenum	Lead	Molybdenum	Aluminum ore (bauxite)
Magnesium	Zinc	Magnesium	Fluorspar
Titanium	Tin	Titanium	Leather hides
Vanadium	Nickel	Bromine	Wool
Bromine	Cobalt	Sulfur	Natural rubber
Sulfur	Mercury	Phosphate rock	
Phosphate rock	Chromium	Cotton	
Cotton	Manganese	Iron Ore	
	Tungsten	Copper	
	Silver	Lead	
	Gold	Zinc	
	Platinum	Nickel	
	Diamonds	Cobalt	
	Aluminum ore (bauxite)	Mercury	
	Asbestos	Chromium	
	Fluorspar	Manganese	
	Potash	Tungsten	
	Petroleum	Silver	
	Natural gas	Gold	
	Leather hides	Platinum	
	Wool	Diamonds	
	Lumber	Asbestos	
	Wood pulp	Potash	
	Natural rubber	Petroleum	
		Natural gas	
		Lumber	
		Wood pulp	

* Industrial raw materials with worldwide consumption greater than $100 million per year are included, except food raw materials.

Note: This list omits some large-volume, low-priced raw materials such as salt, sand, clay, limestone, crushed stone, and the like, in which most large countries are self-sufficient and that do not enter extensively into international trade.

Soviet Union, although in the coming decades the Soviets may begin to put more emphasis on consumer goods.

Soviet industrial production is still largely concentrated on production of basic materials—steel, cement, petroleum, and fertilizer, for example—capital goods, and military equipment. Soviet production of some basic products is nearly as high as that of the U. S. and will probably exceed the U. S. in some products within the next 10 years. If the estimates for production of basic materials are anywhere near right, by 1980 the Soviet Union will be substantially ahead of the U. S. in production of steel, cement, and coal, about even in petroleum, but still lagging in natural gas. The U. S. will probably still be ahead in production of nonferrous metals as a group, but not by very much. . . .

Industrial Power

Steel, cement, coal, petroleum, and other such basic products are the "sinews of industrial power" that constitute the real foundation of a country's economic power. Luxury consumer products do not add to a country's economic power and, in fact, probably detract from a country's economic power in the context of competition between two giants such as the U. S. and U. S. S. R.

The big difference between Soviet and U. S. industrial production, though, is that Soviet production is solidly based on domestic raw materials whereas the U. S. is having to depend more and more on raw materials imported from all over the world.

I believe this situation adds up to a real potential danger to the U. S. The Soviet Union has had world domination as its primary goal since 1917, and the penultimate step in achieving that goal is to surpass the U. S. in economic, political, and military power. The U. S. has been the world's number one power in every sense of the word since 1918. The Soviet Union seeks to displace the U. S. from that position and be recognized as the world's number one power. At present our gross national product is nearly double that of the Soviet Union, but in basic industrial power (steel, petroleum, cement, and the like) the two countries are nearly even. The Soviet Union has been gaining on the U. S. steadily ever since the end of World War II. The Soviet Union's increasingly predominant position in raw materials is a major factor in swinging the balance in favor of the Soviet Union.

Once the Soviet Union believes it is beginning to pull ahead of the U. S. in "basic industrial power," and believing that military and political power follow industrial power, it may begin to put political pressure on suppliers of raw materials to reduce or stop shipments to the U. S. and/or raise prices of raw materials to us and our allies. Countries that might conceivably be pressured in this way include Libya, Zambia, Congo, Nigeria, Gabon, Kenya, Turkey, Iran, Indonesia, Thailand, Philippines, Finland, Sweden, Greece, Yugoslavia, and perhaps even South American countries.

If such a stratagem succeeded even in part, we would be faced with worsening terms of trade in raw materials and a very difficult problem of obtaining enough raw materials to supply our industries. If the Soviets could engender even a partial interdiction of the flow of raw materials from other continents to the U. S., the effect on our industrial production would be disastrous.

Counteraction

To counteract such a possible sequence of events, we should take all possible economic, diplomatic, and legislative steps to strengthen our relationships with key areas now supplying raw materials to us.

There are at least five things I think we should do:

1. Continue to cultivate the best possible relationships with our two neighbors, Canada and Mexico.
2. Continue to cultivate the best possible relationships with Australia and New Zealand.
3. Continue to maintain good relations in Europe, particularly with the non-Communist countries and with Yugoslavia, since Europe is still a significant source of some industrial raw materials.
4. Greatly improve our relationships with all South American countries.
5. Develop much closer ties with the countries of western and southern Africa, from Morocco to South Africa.

Canada and Australia are of the highest importance, with Mexico and New Zealand of lesser importance, as raw material sources. Canada and Australia both have enormous untapped mineral resources and probably still-undiscovered resources.

South America is a vast storehouse of raw materials, and it is essential that we take steps to ensure our continued access to the raw materials of this great continent. Toward that end I urge that the U. S. Government and U. S. private organizations develop new philosophies and new programs that will strengthen the economic, political, and cultural ties between the U. S. and the Latin American countries.

Africa is probably the greatest remaining storehouse of raw materials, and I believe we should develop much closer economic, political, and cultural ties with the countries of western and southern Africa than now exist. Many of the countries of western and southern Africa are important sources of industrial raw materials for the U. S., notably Morocco, Liberia, Nigeria, Gabon, Congo (Kinshasa), Zambia, Rhodesia, Angola, Southwest Africa, and Republic of South Africa. Western Africa is now becoming an important source of petroleum as well as many minerals, notably Nigeria and Angolia (including Cabinda). Eastern Africa is of very little importance as a source of raw materials with some minor exceptions, such as tantalum ore from Mozambique and beryllium ore from Uganda. Also, it is farther away. Northern Africa, except Morocco, seems likely to be dominated by the Soviet Union within a few years.

Asia Least Important

Asia, although the largest continent, is of least importance to us as a source of industrial raw materials, and it is far away. We could get along without raw materials from Asia altogether, if we had adequate access to raw materials from Canada, Australia, Mexico, South America, and western and southern Africa. It is true we now get tin and natural rubber from Indonesia, Malaysia, and Thailand; manganese, mica, and jute from India; copper, copra, and hemp from the Philippines; and a little petroleum from the Persian Gulf, but we could get along without any of these raw materials from Asia. Turkey, an important source of chromium, is a special problem in western Asia since it is a member of NATO but is rapidly being outflanked by the Soviet Union. Also, tin might be a minor problem, but we could probably get along with Western Hemisphere and western African sources of tin plus the further development of substitutes for tin.

The really important industrial raw materials are petroleum, natural gas, coal, iron ore, uranium, copper, lead, zinc, nickel, manganese, chromium, cobalt, and a few others; at the present time we have access to adequate supplies of all these raw materials without relying on Asia. The one big exception to this generalization is our probable need for petroleum from the Persian Gulf in the near future. Persian Gulf oil is the one raw material from the Asian continent that the U. S. may need—and need very badly—within the next decade.

Persian Gulf

Petroleum is one of the really critical raw materials in any country and especially in the more industrialized countries. The Persian Gulf is now the largest petroleum-producing area, and it has by far the largest known reserves of petroleum. At the present time we get very little petroleum from the Persian Gulf (about 2% of U. S. consumption), and we could probably get along for several more years without getting any from that area. We now get enough petroleum to meet our needs from domestic production, Canada, Venezuela, and Trinidad, plus smaller amounts from the Persian Gulf, Libya, and Indonesia. Most petroleum economists agree, though, that by 1980 the U. S. will need to get a lot of petroleum from the Persian Gulf, possibly as much as a third of our requirements according to some analysts, when our total requirements will probably be about 1 billion metric tons compared to 700 million metric tons last year.

By 1980 many other countries in the world also will be depending on the Persian Gulf for a major portion of their petroleum requirements, including western Europe, eastern Europe, Japan, China, other Asian countries, and possibly even the Soviet Union. Most of the Persian Gulf production now goes to western Europe and Japan with smaller amounts to other Asian countries and eastern Africa. Production of petroleum is increasing rapidly in northern and western Africa, and these developments may lessen the future dependence of the U. S. and Europe on the Persian Gulf area. Also, the new petroleum developments in Alaska and in the North Sea will help to lessen the future dependence of the U. S. and Europe on the Persian Gulf area. Nonetheless, barring some very large new discoveries, in 20 or 30 years most of the petroleum left in the world will lie beneath the sands of the deserts surrounding the Persian Gulf.

However, if the price of crude oil from foreign sources should increase substantially as a result of Soviet pressure in the Middle East and North Africa, or for any other reason, new higher cost sources of hydrocarbons would come into play, such as oil shale, tar sands, certain "very heavy crudes" (not normally included in petroleum reserve estimates) and, of course, hydrocarbons from coal. Known deposits of oil shale, tar sands, and very heavy crudes in the Western Hemisphere alone contain hydrocarbons amounting to many times the total known petroleum reserves in the world today.

Research's Role

In addition to making our principal sources of raw materials more secure, we should accelerate research in three very important areas. First, we need to find methods of using lower grades of minerals. As high-grade mineral deposits are depleted, the U. S. and other industrialized countries will be forced to use lower and lower grade minerals. New technologies are being developed continuously to use lower grade minerals, but research in this direction needs to be accelerated.

In countries as large as the U. S., Canada, and Mexico there are undoubtedly large mineral deposits that haven't been discovered. New, more sophisticated techniques of discovery need to be developed to expand North American sources of minerals. Finally, greater recycling of all nondegradable materials is going to become imperative both to preserve our environment and as a means of slowing the drain on our mineral and forest resources. Research in this area involves studies on economic, social, and political factors as well as technical research and development.

Henry H. Villard: Economic Implications for Consumption of 3 Percent Growth

May I start with one general warning: I take it that we are primarily concerned with the broad implications of continued growth. I therefore do not intend to devote time to statistical refinements or the obvious qualifications that should accompany any effort to make comparisons over a long period of time. Instead I propose simply to consider what the shape of our economy would be if the growth rate we achieved from 1865 to 1965 were maintained until 2065.

My best estimate is that over the last century hours worked declined from 67 to 40, or by 40 percent, and the number of years worked from 55 to 45, or 18 percent, while real income per worker increased fivefold, per person sixfold, and per man-hour eightfold.[1] Note that these estimates suggest a growth in real income of considerably less than 3 percent. Even the eightfold increase in real income per man-hour involves only a 2 percent rate of growth. The 3 percent figure I believe stems from, and somewhat underestimates, the increase in total production including the increase in those at work. . . .

Let me concentrate on real income per worker. In 1965 the figure was around $7,200; five times $7,200 is $36,000.* The question then becomes how viable would an economy be which had an average income of $36,000 per worker. To avoid repeating the figure endlessly, let us describe those with $36,000 today as "rich" and those with $36,000 in 2065 as "average." Assuming no change in our present skewed distribution of income, this means that in 2065 one-third of income recipients would have more than $36,000 and two-thirds less.

Obviously a fair number of people, including some even in the academic world, have at present no difficulty in spending $36,000. If one, why not all? Of course one of the things that aids the rich in spending their income is that more of it goes for taxes. Admittedly the overall progressiveness of our tax structure is not great, but I suspect that with present

---

Reprinted from *American Economic Review,* Vol. LVIII, No. 2, May 1968. Reprinted by permission of the American Economic Association and the author.

[1] These estimates are derived from the data presented in Tables 28–1 and 28–2 of my *Economic Performance* (Holt, Rinehart, and Winston, 1962).

* [In rough order of magnitude this is equivalent to a $7–10 trillion GNP. Eds.]

taxes a fivefold expansion of pre-tax income from $7,200 to $36,000 would lead to not more than a fourfold increase in disposable income.

Whether the same relationship will apply in the future—so that by 2065 average disposable income will have increased only fourfold—depends on what happens to government spending. I believe that the nonmilitary share of the national income going to the government will inevitably increase over the next century—primarily because, in substantial contrast to the private economy, most government services appear to be subject to at best constant and more generally diminishing returns. Garbage and sewerage removal, water supply, transportation, even police and fire protection, to say nothing of education, are likely to cost relatively more in the increasingly urban world implied by continued growth. And I shall shortly argue that growth itself will also inevitably raise problems which will require a further expansion of government activity.

Total government spending, on the other hand, depends to a large extent on what happens to military spending. One can argue that, if we reach 2065, peace must have been maintained and, as a result, military spending must have decreased. Perhaps so. But I shall argue toward the end of this paper that continuing peace is unlikely unless decreases in military spending are accompanied by substantial increases in foreign aid. In fact it would not surprise me if the government's need for funds for both domestic and foreign programs were to prove so massive that the increase in disposable income over the next century turned out to be less than fourfold. Only if peace is possible without foreign aid—if we can exist in splendid isolation—is the increase likely, as I see it, to be much more than fourfold.

Thus I do not think we will be wide of the mark if we take as our basic question: does a rich man with $36,000 today spend his income in ways which would be impossible, or meaningless, or yield him very much less satisfaction if all incomes averaged $36,000? Note that I am not concerned with the problem of invidious comparisons: as I assume no change in income distribution, I propose simply to assume that the $180,000 man in 2065 will still get the same pleasure from feeling superior to the $36,000 man that a $36,000 man today gets from feeling superior to the $7,200 man. Our basic question has two aspects: production and consumption. From the point of view of production it may be quite impossible to provide the average person in 2065 with exactly the same items as the rich received in 1965. Specifically, if 1 percent of beef production is filet mignon and it were exclusively eaten by the rich in 1965, it would obviously be impossible—short of an unlikely reengineering of the steer or an immensely wasteful production of meat—to provide the average consumer with filet mignon in 2065. But the rich in 1865 had venison, buffalo meat, and passenger pigeon eggs on a scale not open to the rich in 1965, so that to the extent that identical items cannot be provided we are faced primarily with the well-known index number problem of measuring equivalents. This we have "solved"—or at least we think we have—or we would not be

able to compare 1865 with 1965. I admit to reservations regarding our "solutions"; I am not clear that, despite all the dedicated work of our national income statisticians, we have been able to establish an entirely satisfactory trade-off between a washing machine and a washerwoman, or a TV dinner and a built-in cook. But I believe that the decline in the quantity of personal services available to the rich between 1865 and 1965 was probably greater than the likely future decline in the quantity of such services available to the rich in 1965 and the average person in 2065. . . .

Let me next turn to the distinction between private and social product. When Pigou developed the concept, I believe he had production primarily in mind. But I want first to apply the distinction to consumption and then later to production. Let me telegraph my punch: I believe that there are significant differences between the private and social benefits of various kinds of additional consumption. Hence, while I am inclined to believe that a world in which incomes average $36,000 would be basically viable, the degree of viability would vary greatly depending on the types of consumption involved. To the extent that this is correct it raises obvious questions regarding the desirability of unrestrained consumer sovereignty and the usefulness of the national income as a measure of welfare, as the national income is primarily the sum of the money value of private rather than social benefits.

Let me illustrate what is involved. While, as just noted, there may be problems on the production side, I see no reason why there should be a divergence between the private and social benefit, which means I see no reason why as a consumer I should be harmed if the average person in 2065 were to have a diet equivalent to the present diet of the rich, for what my neighbor eats does not adversely affect me. But, even if there is no production problem in providing the average man with the cars in general and the Cadillacs in particular now consumed by the rich, the provision of such additional cars would have a severely adverse impact on him as a result of more crowded highways, more accidents, more difficult parking, and more smog. I think the basic distinction is that the consumption of more cars affects the general environment—typically adversely—while the consumption of more food for practical purposes does not. Let us call consumption which does not affect the environment "neutral," that which harms it "adverse," and that which improves it "beneficial."

I suppose I am optimistic regarding the viability of large increases in consumption because most consumption strikes me as neutral while some is surely beneficial. Let me repeat that I am throughout assuming that the production problem has been solved—at least in terms of equivalents—by the very growth of output which we have premised. Thus the issue is not redistribution, for example, of essentially limited medical services from the rich to the average person as a result of health insurance, but rather whether there would be any repercussions on the average consumer in 2065 if there were enough medical services available in that year to serve the average consumer as well as the rich were served in 1965. In this case the additional consumption, to the extent that it reduced contagious diseases, would clearly fall into the beneficial category. Moreover, even if the appeal of a fashionable restaurant is that it is hard to get into, there is no reason why such a restaurant a century hence should not charge prices which would make it as hard for a $36,000 man to visit as it is today for a $7,200 a year man to eat regularly at the Four Seasons. As a result, it seems to me that average spending on such important items as food, clothing, personal care, medical care, personal business, and even household operation could be increased four- or five-fold and still remain neutral. Yet among them these categories account for two-thirds of consumption expenditures.

Housing, amounting to 13 percent of consumption, is less clear cut. We need to distinguish between floor space and land space. In a typical suburb today a rich man has perhaps three times as much of both as the average person. Providing three times as much floor space should be easy if we are willing to expand upward, but providing three times as much land space has major implications for the transportation system and urban sprawl. Much depends, therefore, on whether we seek to provide the average man in 2065 with the land space today consumed by the rich willing to live on Park Avenue at a land density greater than a Harlem slum or with the space consumed by those rich living in the suburbs at a fraction of the average density.

Transportation, amounting to 12 percent of consumption, overwhelmingly represents expenditures on cars. For anyone who wants to understand the impact of the car on our civilization I recommend a trip by car to the Soviet Union, where as late as 1964 it was possible for a casual tourist to park overnight directly in front of the best hotel in Moscow. Certainly the most obvious difference between our two economies is not such matters as the ownership of the means of production but the simple fact that we have cars and they do not! The only encouraging factor that I can see in this area is that today cars registered are equal to 60 percent of persons eighteen years and over. The sort of increase in incomes we are considering would probably involve at least a doubling in registrations per person and perhaps a threefold increase in road use, and therefore congestion, as jalopies and Volkswagens are upgraded to Cadillacs and mileage driven increases. But even the increase in use is likely to be considerably less than the increase in real income.

Recreation is the final category worth special discussion. Most of the spending involved—for example, purchase of books and magazines—is in my terminology neutral. But much spending is for vacations in which scenery is of the essence, be it the view from a summer home, in a national park, or from a cruising boat. To a major extent the problems just discussed in regard to housing and transportation stem from the fact that land is limited in quantity, but productive effort can in both cases appreciably mitigate the adverse repercussions on the environment. This is much harder to do in the case of scenery, which is, so to speak, land-intensive. A national park, for example, which is so jammed that one has to line up for a day or so before being allowed in is hardly an ideal example of the virgin wilderness it is supposed to illustrate. Again it has been estimated that if every British family were to want a bit of sea front, the entire sea coast of England, Scotland, and Wales would provide an average per family of 33 front feet. Rising incomes will certainly make solitude and unspoiled scenery very hard to come by.

To sum up thus far: even a fivefold increase in per capita real income and consumption spending over the next century does not seem to me likely to result in serious problems. Admittedly our individualistic tradition from frontier days makes us feel that a man's right to his property includes the right to create an eyesore. It was, for example, only with considerable difficulty that an ordinance was recently passed on Martha's Vineyard denying a man the right to have more than two unlicensed cars—i.e., wrecks—in his front yard. But if we come to be willing to use part of the increase in real income to combat the repercussions on the environment of adverse consumption, it should be possible to keep such repercussions within reasonable bounds. Suppose, for example, we were to agree to use part of the growth of income to cleanse the landscape of dead cars. Once agreement had been reached on the objective, I do not believe it would be difficult to work out effective mechanisms even though it might involve abridging the right of car owners to dump a couple of tons of trash anywhere they wanted. Note, moreover, that the more we accept a public obligation to use the growth of income to preserve the environment, the less will be the growth of disposable income and, therefore, other things equal, the increase in adverse consumption.

Nor am I much less optimistic even when account is taken of the adverse repercussions on the environment as a result of production. Admittedly Pigou distinguished between private and social products before the problem had become particularly acute in the United States. Certainly business has felt as free to dump trash as consumers: unless there is clear scrap value in an item or an alternative use for the land in question, barges are left to rot on the first available mud flat, piers to fall into the sea, and factories to deteriorate into eyesores. But because we are a big country, we could afford to foul large areas of our environment and still have much left that was unspoiled.

Basically, what we need to do is to devise ways of determining the socially desirable cost of production, including the cost of removing the debris of production. Overall, I think it is fair to say that we have made very little progress in meeting this need. In fact, I am not clear that the problem is widely recognized by economists. Certainly, if I may use what I consider the best single criterion of recognition, it has not yet gotten into Samuelson's text, which tends to treat private costs as if they were all that mattered. In sharp contrast, I submit, to take a specific example, that the construction costs incurred by Con Edison are altogether too important a matter to be left to a private utility. So little has this been generally recognized that it recently took a knock-down, drag-out fight to obtain a court order requiring the Federal Power Commission to give significant consideration to the adverse effects on the environment of the various ways in which power can be generated and transmitted. Again, despite our wealth, we today permit coal mining companies to create immense wastelands in the course of strip mining. But Jamaica, despite her very much lower level of living, has insisted that bauxite mines replace their divots; in fact, because gradients are reduced, the restored land is in many cases more suitable for agriculture than before the mining took place.

Once economists get over considering private costs as sacred, I believe that they have a major contribution to make in this area. For if utility costs are too important to be left to private utilities, they are also too important to be left to conservationists. What we need to determine is the amount to be added to private costs to achieve the socially desirable reduction in the adverse repercussions on the environment of the activity in question—an amount far easier to describe than to determine. For in many cases costs will be hard to measure in dollars and cents. How many million people, for example, is it proper to disturb in order that a handful riding on a supersonic transport may get to their destinations a couple of hours earlier? Unless economists can develop techniques for measuring, in some fashion or other, such almost unmeasurable things, decisions in the future will increasingly be made at best by adversary proceedings in our law courts and at worst by plain power politics.[2]

To sum up once more: I do not believe that the adverse repercussions on the environment resulting from both the increase in consumption and in production involved in even a fivefold increase in real income per capita are likely to be unacceptably large, especially if we devote a significant part of the increase in real income to a reduction of the adverse repercussions in question.

I become substantially less optimistic when consideration is given to probable population growth. For our most recent rate of growth, if maintained for a century, would result in an approximately threefold increase in our numbers, to perhaps 600 millions in 2065. Our average density, however, would even then be appreciably less than present English density. Hence I do not see any basis for arguing that a threefold increase in our numbers would be particularly difficult to achieve.

But if it were to occur, it would immensely intensify the problems raised by both adverse consumption and production. Specifically, if our estimates are correct that a four- or fivefold increase in real income will be

---

[2] Stephen Enke deserves commendation for his pioneering effort to apply cost-benefit analysis to the supersonic transport in the May, 1967, *A.E.R.*

likely to result in no more than a doubling of automobile registrations to 150 millions, then we could probably live with the problem without too serious difficulties. But if registrations rose to 450 millions as a result of a threefold increase in population, then even to park the cars bumper to bumper in dead storage would require half the land area of Connecticut. Again, if the average person in 2065 opted to live like our present suburban rich with three times our present average land space, then nine times as much land would be required. In short, as an increase in average real income per capita appears inevitable, it seems to me that whether we face over the next century problems which will be relatively easy to solve or basic changes in the sort of world in which we live depends primarily on our rate of population growth over the period. . . .

I suspect what really lies behind my assignment today is the question, "Is growth necessary?" I trust my discussion has made clear that the urgency of growth depends on the sort of world we envisage. Even if there were no further growth in our population and we could live isolated from the problems of the rest of the world, I would still be inclined to favor a rate of growth of real income at least as rapid as we achieved over the last century, partly because I have difficulty seeing why the average American should be asked to wait more than a century to live as well as I do today. But I would have to admit that the growth under such circumstances would not seem to me desperately urgent. I would also admit that, the more rapid the growth of income, the more rapid would be the growth of consumption and—other things equal—the growth of adverse consumption. But, as we have seen, the repercussions of adverse consumption can be mitigated, or even eliminated, by public spending, and rapid growth of income should make it easier to obtain increases in public spending. I can in fact conceive that slow growth might intensify our difficulties by making us quite unwilling to devote to the public sector the resources that are needed. For I believe that only when disposable income is rising rapidly are we likely to accept what seems to me to be essential increases in the public sector. I offer the thought, therefore, that there may well be a substantial range over which changes in the rate of growth of real income leave the problem of adjusting to growth substantially unchanged. To the extent that this proves to be correct, then I see no reason for not trying to achieve a fairly rapid growth in our per capita income even in the absence of population growth. Certainly I can see little prospect for the success, in the absence of fairly rapid growth, of Galbraith's effort to expand the public sector by simply telling us that we are affluent.

But, whether we like it or not, population growth in the United States is not going to end immediately, and the larger our population, the more urgent does rapid growth in real per capita income become. This is so because an increase in real income resulting from population growth is likely to cause a significantly more rapid increase in adverse consumption than an equivalent increase in real income resulting from higher income per capita. The relative increase reflects primarily the fact that land is fixed, so that, as population increases, we will have to devote increasing amounts to improving and "reclaiming" land. What this means on a less general level of abstraction is that, if our numbers are going to increase to even 300 millions, we are not going to be able to continue to house the increase in population in suburbs built on land previously in agricultural use miles from the central city but will instead have to undertake massive expenditures on urban renewal. And I hardly need add that immense investment in improving our transportation system will also be needed.

In short, it seems to me that, in view of our probable population growth continued rapid growth of per capita real income is highly desirable even when we view the matter from our own point of view considered apart from any obligations we may have to the rest of the world. But, to the extent that we as the richest nation the world has ever known do in fact accept a significant obligation to concern ourselves with the problems of the rest of the world, then the growth required becomes simply astronomical in size. . . .

4 Precedents and Prospects

For each claim that economic growth and increased abundance are the foundations of the coming millenium, there is offered another claim like Wordsworth's, that it is "a sordid boon."

> In the [Hugh] Hefner vocabulary, "materialism isn't—and shouldn't be—a dirty word." You shouldn't feel guilty about earning money and spending it, he says, and when you decide to build the world's "wildest and most exciting" private plane, an extra million here or there doesn't make that much difference. . . . His objective is to set a high standard of comforts, and bring the rest of the world up to it. "There is never a reason to give up living well" (*Look,* June 2, 1970).

> The world is too much with us; late and soon,
> Getting and spending, we lay waste our powers:
> Little we see in Nature that is ours;
> We have given our hearts away, a sordid boon!
> The sea that bares her bosom to the moon;
> The winds that will be howling at all hours,
> And are up-gather'd now, like sleeping flowers;
> For this, for everything, we are out of tune;
> It moves us not.—Great God! I'd rather be
> A Pagan suckled in a creed outworn;
> So might I, standing on this pleasant lea,
> Have glimpses that would make me less forlorn;
> Have sight of Proteus rising from the sea;
> Or hear old Triton blow his wreathed horn.
>
> "The World"
> William Wordsworth, 1770–1850

For the last 200 years of the Industrial Revolution, the heady optimism of "progress" has been unstoppable; we were overcoming mankind's age-old plagues of poverty, back-breaking labor, and disease. But today even the growthmen are becoming apologetic for the unanticipated by-products of economic growth and less willing to equate increments in GNP with increased human welfare. Others are taking a strong personal, and sometimes political, stance against alienating work, bureaucratic efficiency, and consumer mentality, all basic components of a growth-oriented society. It is important to realize, however, that this phenomenon is historically unprecedented only insofar as it constitutes a large-scale social movement. The past contains numerous examples of individual protest against the clamor of "advancing civilization."

One of these is Henry David Thoreau, whose *Walden; Or, Life in the Woods* is probably more pertinent in today's world than the world of 1845–1847, when Thoreau spent his two years at Walden Pond to build his cabin, to till his beanfield, and "to live deliberately, to front only the essential facts of life. . . ." Thoreau sounds very contemporary in his condemnation of voluntary slavery in the pursuit of possessions and of the technology in which "a few are riding but the rest are run over." His quest was for the economy of essential living, for freedom, and "the poverty that enjoys true wealth."

It is also comforting to have perhaps the greatest economist of this century say "the economic problem is not . . . the permanent problem of the human race." John Maynard Keynes—mathematician, financier, and author of *The General Theory of Employment, Interest, and Money*—was also the friend of George Bernard Shaw and Pablo Picasso, patron of the ballet, connoisseur of art, and a central figure in London's brilliant Bloomsbury Group of writers and intellectuals. In his essay "Economic Possibilities for Our Grandchildren," written in the dark depression days of the 1930s, he could say, "I see us free, therefore, to return to some of the most sure and certain principles of religion and traditional virtue—that avarice is a vice, that the exaction of usury is a misdemeanour, and the love of money is detestable, that those walk most truly in the paths of virtue and sane wisdom who take least thought for the morrow."

As mankind approaches the twenty-first century and moves closer to an indeterminant ecological point of no return, increasing numbers of the young, literally Keynes' grandchildren, are calling for the economy Keynes foresaw in his essay. Economic "science" today is based on exchange, on *quid pro quo* in social relations, on possession and accumulation of material objects. Thus, economics, traditional life styles, and life-denying social institutions are being rejected in favor of new ways that affirm life. Increasing numbers of people are dropping out of that economic mainstream of production and consumption. In addition to isolated individuals and communes, one can observe youth communities burgeoning forth in almost every major city and many minor ones, as well as in rural areas. These are intended to be liberated zones where people can be free to transform their own lives, their relations with others, to live in harmony with nature, and eventually to be able to move by action as well as example to bring change to the wider society. The most noteworthy of these communities is in Berkeley, and the "Berkeley Liberation Program," which came out of the crowd rushes to the depot, and the conductor shouts "All aboard!" when the smoke is blown away and the vapor condensed, it will be perceived that a few are riding, but the rest are run over,—and it will be called, and will be, "A melancholy accident." . . .

A lady once offered me a mat, but as I had no room to spare within the house, nor time to spare within or without to shake it, I declined it, preferring to wipe my feet on the sod before my door. It is best to avoid the beginnings of evil. . . .

For more than five years I maintained myself thus solely by the labor of my hands, and I found that, by working about six weeks in a year, I could meet all the expenses of living. The whole of my winters, as well as most of my summers, I had free and clear for study. I have thoroughly tried schoolkeeping, and found that my expenses were in proportion, or rather out of proportion, to my income, for I was obliged to dress and train, not to say think and believe, accordingly, and I lost my time into the bargain. As I did not teach for the good of my fellow-men, but simply for a livelihood, this was a failure. I have tried trade; but I found that it would take ten years to get under way in that, and that then I should probably be on my way to the devil. I was actually afraid that I might by that time be doing what is called a good business. When formerly I was looking about to see what I could do for a living, some sad experience in conforming to the wishes of friends being fresh in my mind to tax my ingenuity, I thought often and seriously of picking huckleberries; that surely I could do, and its small profits might suffice,—for my greatest skill has been to want but little,—so little capital it required, so little distraction from my wonted moods, I foolishly thought. While my acquaintances went unhesitatingly into trade or the professions, I contemplated this occupation as most like theirs; ranging the hills all summer to pick the berries which came in my way, and thereafter carelessly dispose of them; so, to keep the flocks of Admetus. I also dreamed that I might gather the wild herbs, or carry evergreens to such villagers as loved to be reminded of the woods, even to the city, by hay-cart loads. But I have since learned that trade curses everything it handles; and though you trade in messages from Heaven, the whole curse of trade attaches to the business. . . .

For myself I found that the occupation of a day-laborer was the most independent of any, especially as it required only thirty or forty days in a year to support one. The laborer's day ends with the going down of the sun, and he is then free to devote himself to his chosen pursuit, independent of his labor; but his employer, who speculates from month to month, has no respite from one end of the year to the other.

In short, I am convinced, both by faith and experience, that to maintain one's self on this earth is not a hardship but a pastime, if we will live simply and wisely; as the pursuits of the simpler nations are still the sports of the more artificial. It is not necessary that a man should earn his living by the sweat of his brow, unless he sweats easier than I do. . . .

I went to the woods because I wished to live deliberately, to front only the essential facts of life, and see if I could not learn what it had to teach, and not, when I came to die, discover that I had not lived. I did not wish to live what was not life, living is so dear; nor did I wish to practise resignation, unless it was quite necessary. I wanted to live deep and suck out all the marrow of life, to live so sturdily and Spartan-like as to put to rout all that was not life, to cut a broad swath and shave close, to drive life into a corner, and reduce it to its lowest terms, and, if it proved to be mean, why then to get the whole and genuine meanness of it, and publish its meanness to the world; or if it were sublime, to know it by experience, and be able to give a true account of it in my next excursion. . . .

What I Lived For

Flint's or Sandy Pond, in Lincoln, our greatest lake and inland sea, lies about a mile east of Walden. It is much larger, being said to contain one hundred and ninety-seven acres, and is more fertile in fish; but it is comparatively shallow, and not remarkably pure. A walk through the woods thither was often my recreation. . . .

The Ponds

Flint's Pond! Such is the poverty of our nomenclature. What right had the unclean and stupid farmer, whose farm abutted on this sky water, whose shores he has ruthlessly laid bare, to give his name to it? Some skin-flint, who loved better the reflecting surface of a dollar, or a bright cent, in which he could see his own brazen face; who regarded even the wild ducks which settled in it as trespassers; his fingers grown into crooked and horny talons from the long habit of grasping harpy-like;—so it is not named for me. I go not there to see him nor to hear of him; who never *saw* it, who never bathed in it, who never loved it, who never protected it, who never spoke a good word for it, nor thanked God that He had made it. Rather let it be named from the fishes that swim in it, the wild fowl or quadrupeds which frequent it, the wild flowers which grow by its shores, or some wild man or child the thread of whose history is interwoven with its own; not from him who could show no title to it but the deed which a like-minded neighbor or legislature gave him,—him who thought only of its money value; whose presence perchance cursed all the shores; who exhausted the land around it, and would fain have exhausted the waters within it; who regretted only that it was not English hay or cranberry meadow,—there was nothing to redeem it, forsooth, in his eyes,—and would have drained and sold it for the mud at its bottom. It did not turn his mill, and it was no

privilege to him to behold it. I respect not his labors, his farm where everything has its price, who would carry the landscape, who would carry his God, to market, if he could get anything for him; who goes to market *for* his god as it is; on whose farm nothing grows free, whose fields bear no crops, whose meadows no flowers, whose trees no fruits, but dollars; who loves not the beauty of his fruits, whose fruits are not ripe for him till they are turned to dollars. Give me the poverty that enjoys true wealth. Farmers are respectable and interesting to me in proportion as they are poor,—poor farmers. A model farm! where the house stands like a fungus in a muck-heap, chambers for men, horses, oxen, and swine, cleansed and uncleansed, all contiguous to one another! Stocked with men! A great grease-spot, redolent of manures and buttermilk! Under a high state of cultivation, being manured with the hearts and brains of men! As if you were to raise your potatoes in the churchyard! Such is a model farm. . . .

Why should we be in such desperate haste to succeed and in such desperate enterprises? If a man does not keep pace with his companions, perhaps it is because he hears a different drummer. Let him step to the music which he hears, however measured or far away. . . .

Conclusion

However mean your life is, meet it and live it; do not shun it and call it hard names. It is not so bad as you are. It looks poorest when you are richest. The faultfinder will find faults even in paradise. Love your life, poor as it is. You may perhaps have some pleasant, thrilling, glorious hours, even in a poorhouse. The setting sun is reflected from the windows of the almshouse as brightly as from the rich man's abode; the snow melts before its door as early in the spring. . . .

John Maynard Keynes: Economic Possibilities for Our Grandchildren

---

Let us, for the sake of argument, suppose that a hundred years hence we are all of us, on the average, eight times better off in the economic sense than we are to-day. Assuredly there need be nothing here to surprise us.

From *Essays in Persuasion,* by John Maynard Keynes. Reprinted by permission of Harcourt Brace Jovanovich, Inc. and Macmillan London and Basingstoke.

Now it is true that the needs of human beings may seem to be insatiable. But they fall into two classes—those needs which are absolute in the sense that we feel them whatever the situation of our fellow human beings may be, and those which are relative in the sense that we feel them only if their satisfaction lifts us above, makes us feel superior to, our fellows. Needs of the second class, those which satisfy the desire for superiority, may indeed be insatiable; for the higher the general level, the higher still are they. But this is not so true of the absolute needs—a point may seen be reached, much sooner perhaps than we are all of us aware of, when these needs are satisfied in the sense that we prefer to devote our further energies to non-economic purposes.

Now for my conclusion, which you will find, I think, to become more and more startling to the imagination the longer you think about it.

I draw the conclusion that, assuming no important wars and no important increase in population, the *economic problem* may be solved, or be at least within sight of solution, within a hundred years. This means that the economic problem is not—if we look into the future—*the permanent problem of the human race*.

Why, you may ask, is this so startling? It is startling because—if, instead of looking into the future, we look into the past—we find that the economic problem, the struggle for subsistence, always has been hitherto the primary, most pressing problem of the human race—not only of the human race, but of the whole of the biological kingdom from the beginnings of life in its most primitive forms.

Thus we have been expressly evolved by nature—with all our impulses and deepest instincts—for the purpose of solving the economic problem. If the economic problem is solved, mankind will be deprived of its traditional purpose.

Will this be a benefit? If one believes at all in the real values of life, the prospect at least opens up the possibility of benefit. Yet I think with dread of the readjustment of the habits and instincts of the ordinary man, bred into him for countless generations, which he may be asked to discard within a few decades.

To use the language of to-day—must we not expect a general "nervous breakdown"? We already have a little experience of what I mean—a nervous breakdown of the sort which is already common enough in England and the United States amongst the wives of the well-to-do classes, unfortunate women, many of them, who have been deprived by their wealth of their traditional tasks and occupations—who cannot find it sufficiently amusing, when deprived of the spur of economic necessity, to cook and clean and mend, yet are quite unable to find anything more amusing.

To those who sweat for their daily bread leisure is a longed-for sweet—until they get it.

There is the traditional epitaph written for herself by the old charwoman:—

> Don't mourn for me, friends, don't weep for me never.
> For I'm going to do nothing for ever and ever.

This was her heaven. Like others who look forward to leisure, she conceived how nice it would be to spend her time listening-in—for there was another couplet which occurred in her poem:—

> With psalms and sweet music the heavens'll be ringing,
> But I shall have nothing to do with the singing.

Yet it will only be for those who have to do with the singing that life will be tolerable—and how few of us can sing!

Thus for the first time since his creation man will be faced with his real, his permanent problem—how to use his freedom from pressing economic cares, how to occupy the leisure, which science and compound interest will have won for him, to live wisely and agreeably and well.

The strenuous purposeful money-makers may carry all of us along with them into the lap of economic abundance. But it will be those peoples, who can keep alive, and cultivate into a fuller perfection, the art of life itself and do not sell themselves for the means of life, who will be able to enjoy the abundance when it comes.

Yet there is no country and no people, I think, who can look forward to the age of leisure and of abundance without a dread. For we have been trained too long to strive and not to enjoy. It is a fearful problem for the ordinary person, with no special talents, to occupy himself, especially if he no longer has roots in the soil or in custom or in the beloved conventions of a traditional society. To judge from the behaviour and the achievements of the wealthy classes to-day in any quarter of the world, the outlook is very depressing! For these are, so to speak, our advance guard—those who are spying out the promised land for the rest of us and pitching their camp there. For they have most of them failed disastrously, so it seems to me—those who have an independent income but no associations or duties or ties—to solve the problem which has been set them.

I feel sure that with a little more experience we shall use the new-found bounty of nature quite differently from the way in which the rich use it to-day, and will map out for ourselves a plan of life quite otherwise than theirs.

For many ages to come the old Adam will be so strong in us that everybody will need to do *some* work if he is to be contented. We shall do more things for ourselves than is usual with the rich to-day, only too glad to have small duties and tasks and routines. But beyond this, we shall endeavour to spread the bread thin on the butter—to make what work there is still to be done to be as widely shared as possible. Three-hour shifts or a fifteen-hour week may put off the problem for a great while. For three hours a day is quite enough to satisfy the old Adam in most of us!

There are changes in other spheres too which we must expect to come. When the accumulation of wealth is no longer of high social importance, there will be great changes in the code of morals. We shall be able to rid ourselves of many of the pseudo-moral principles which have hag-ridden us for two hundred years, by which we have exalted some of the most distasteful of human qualities into the position of the highest virtues. We shall be able to afford to dare to assess the money-motive at its true value. The love of money as a possession—as distinguished from the love of money as a means to the enjoyments and realities of life—will be recognised for what it is, a somewhat disgusting morbidity, one of those semi-criminal, semi-pathological propensities which one hands over with a shudder to the specialists in mental disease. All kinds of social customs and economic practices, affecting the distribution of wealth and of economic rewards and penalties, which we now maintain at all costs, however distasteful and unjust they may be in themselves, because they are tremendously useful in promoting the accumulation of capital, we shall then be free, at last, to discard.

Of course there will still be many people with intense, unsatisfied purposiveness who will blindly pursue wealth—unless they can find some plausible substitute. But the rest of us will no longer be under any obligation to applaud and encourage them. For we shall inquire more curiously than is safe to-day into the true character of this "purposiveness" with which in varying degrees Nature has endowed almost all of us. For purposiveness means that we are more concerned with the remote future results of our actions than with their own quality or their immediate effects on our own environment. The "purposive" man is always trying to secure a spurious and delusive immortality for his acts by pushing his interest in them forward into time. He does not love his cat, but his cat's kittens; nor, in truth, the kittens, but only the kittens' kittens, and so on forward for ever to the end of cat-dom. . . .

I see us free, therefore, to return to some of the most sure and certain principles of religion and traditional virtue—that avarice is a vice, that the exaction of usury is a misdemeanour, and the love of money is detestable, that those walk most truly in the paths of virtue and sane wisdom who take least thought for the morrow. We shall once more value ends above means and prefer the good to the useful. We shall honour those who can teach us how to pluck the hour and the day virtuously and well, the delightful people who are capable of taking direct enjoyment in things, the lilies of the field who toil not, neither do they spin.

But beware! The time for all this is not yet. For at least another hundred years we must pretend to ourselves and to every one that fair is

foul and foul is fair; for foul is useful and fair is not. Avarice and usury and precaution must be our gods for a little longer still. For only they can lead us out of the tunnel of economic necessity into daylight.

I look forward, therefore, in days not so very remote, to the greatest change which has ever occurred in the material environment of life for human beings in the aggregate. But, of course, it will all happen gradually, not as a catastrophe. Indeed, it has already begun. The course of affairs will simply be that there will be ever larger and larger classes and groups of people from whom problems of economic necessity have been practically removed. The critical difference will be realised when this condition has become so general that the nature of one's duty to one's neighbour is changed. For it will remain reasonable to be economically purposive for others after it has ceased to be reasonable for oneself.

The *pace* at which we can reach our destination of economic bliss will be governed by four things—our power to control population, our determination to avoid wars and civil dissensions, our willingness to entrust to science the direction of those matters which are properly the concern of science, and the rate of accumulation as fixed by the margin between our production and our consumption; of which the last will easily look after itself, given the first three.

Meanwhile there will be no harm in making mild preparations for our destiny, in encouraging, and experimenting in, the arts of life as well as the activities of purpose.

But, chiefly, do not let us overestimate the importance of the economic problem, or sacrifice to its supposed necessities other matters of greater and more permanent significance. It should be a matter for specialists—like dentistry. If economists could manage to get themselves thought of as humble, competent people, on a level with dentists, that would be splendid!

The Berkeley Liberation Program

Power to the Imagination—All Power to the People

The people of Berkeley passionately desire human solidarity, cultural freedom, and peace.

Berkeley is becoming a revolutionary example throughout the world. We are now under severe attack by the demons of despair, ugliness and

---

From *Leviathan,* Summer 1969. Reprinted by permission of the publisher.

fascism. We are being strangled by reactionary powers from Washington to Sacramento.

Our survival depends on our ability to overcome past inadequacies and to expand the revolution. We have not done enough to build a movement that is both personally humane and politically radical.

The people of Berkeley must increase their combativeness; develop, tighten and toughen their organizations; and transcend their middle-class, ego-centered life styles. We shall resist our oppressors by establishing a zone of struggle and liberation, and of necessity shall defend it. We shall create a genuine community and control it to serve our material and spiritual needs. We shall develop new forms of democratic participation and new, more humane styles of work and play. In solidarity with other revolutionary centers and movements, our Berkeley will permanently challenge the present system and act as one of many training grounds for the liberation of the planet.

1. We will make Telegraph Avenue and the South Campus a strategic free territory for revolution. Historically this area is the home of political radicalism and cultural revolution. We will resist plans to destroy the South Campus through University-business expansion and pig assaults. We will create malls, parks, cafes and places for music and wandering. Young people leaving their parents will be welcome with full status as members of our community. Businesses on the Avenue should serve the humanist revolution by contributing their profits to the community. We will establish cooperative stores of our own, and combine them within an Avenue cooperative.

2. We will create our revolutionary culture everywhere. Everyone should be able to express and develop himself through art—work, dance, sculpture, gardening and all means open to the imagination. Materials will be made available to all people. We will defy all puritanical restraints on culture and sex. We shall have media—newspapers, posters and leaflets, radio, TV, films and skywriting—to express our revolutionary community. We will stop the defiling of the earth; our relation to nature will be guided by reason and beauty rather than profit. The civilization of concrete and plastic will be broken and natural things respected. We shall set up urban and rural communes where people can meet for expression and communication. Many Berkeley streets bear little traffic and can be grassed over and turned into people's parks. Parking meters will be abolished and we will close areas of downtown and South Campus to automotive traffic. We shall celebrate the holidays of liberation with fierce dancing.

3. We will turn the schools into training grounds for liberation. Beneath the progressive facade of Berkeley's schools, students continue to be regimented into accepting the existing system. The widely-celebrated integration of the schools is nothing in itself, and only perpetuates many illusions of white liberalism. The basic issue is creating an educational system in which students have real power and which prepares the young to participate in a revolutionary world. Students must destroy the senile dictatorship of adult teachers and bureaucrats. Grading, tests, tracking, demotions, detentions and expulsions must be abolished. Pigs and narcs have no place in a people's school. We will eliminate the brainwashing, fingernail-cutting mass production of junior cogs for tight-ass America's old age home war machine. Students will establish independent educational forms to create revolutionary consciousness while continuing to struggle for change in the schools.

4. We will destroy the university unless it serves the people. The University of California is not only the major oppressive institution in Berkeley, but a major brain center for world domination. UC attempts to kill radical politics and culture in Berkeley while it trains robots for corporations and mental soldiers to crush opposition from Delano to Vietnam.

Students should not recognize the false authority of the regents, administration and faculty. All students have the right to learn what they want, from whom they want, and in the manner they decide; and the right to take political action without academic penalty. We will build a movement to make the University relevant to the Third World, workers, women and young people searching for human values and vocations. Our battles will be conducted in the classrooms and the streets.

We will shatter the myth that UC is a sacred intellectual institution with a special right to exist. We will change this deadly Machine which steals our land and rapes our minds, or we will stop its functioning. Education can only begin when we're willing to close the University for what we believe.

5. We will struggle for the full liberation of women as a necessary part of the revolutionary process. While the material oppression of women varies in different classes, male supremacy pervades all social classes. We will resist this ideology and practice which oppresses all women. As we struggle to liberate ourselves, many of the problems of inequality, authoritarianism and male chauvinism in the Berkeley movement will be overcome.

We will create an unfettered identity for women. We will abolish the stifling masculine and feminine roles that this society forces on us all. Women will no longer be defined in terms of others than themselves—by their relationships to men and children. Likewise, men will not be defined by their jobs or their distorted role as provider. We seek to develop whole human beings and to bring together the most free and beautiful aspects of women and men.

We will end the economic oppression of women: job discrimination, the manipulation of women as consumers, and media exploitation of women as sexual objects.

We demand the full control of our own bodies and towards that end will establish free birth control and abortion clinics. We will choose our own sexual partners; we will eliminate the demeaning hustling scene in Berkeley which results from male chauvinism and false competition among men and among women. We will not tolerate harrassment in the parks, streets, and public places of Berkeley.

We will resist all false concepts of chivalry and protectiveness. We will develop self-reliance and the skills of self defense. We will establish female communes so that women who so choose can have this free space to develop themselves as human beings.

We will end all forms of male supremacy by ANY MEANS NECESSARY!

6. We will take communal responsibility for basic human needs. High-quality medical and dental care, including laboratory tests, hospitalization, surgery and medicines will be made freely available. Child-care collectives staffed by both men and women, and centers for the care of strung-out souls, the old and the infirm will be established. Free legal services will be expanded. Survival needs such as crash pads, free transportation, switchboards, free phones and free food will be met.

7. We will protect and expand our drug culture. We relate to the liberating potential of drugs for both the mind and the body politic. Drugs inspire us to new possibilities in life which can only be realized in revolutionary action. We intend to establish a drug distribution center and a marijuana cooperative.

As a loving community we shall establish drug information centers and free clinics. We will resist the enforcement of all drug laws in our community. We will protect people from narcs and burn artists. All drug busts will be defined as political and we will develop all necessary defense for those arrested.

8. We will break the power of the landlords and provide beautiful housing for everyone. Through rent strikes, direct seizures of property and other resistance campaigns, the large landlords, banks and developers who are gouging higher rents and spreading ugliness will be driven out. We shall force them to transfer housing control to the community, making decent housing available according to people's needs. Coordinated housing councils will be formed on a neighborhood basis to take responsibility for rents and building conditions. The housing councils will work with architects to plan for a beautiful community. Space will be opened up and living communes and revolutionary families will be encouraged.

9. We will tax the corporations, not the working people. The people cannot tolerate escalating taxes which are wasted in policing the world while businessmen are permitted to expand their profits in the midst of desperate social need. Berkeley cannot be changed without confronting the industries, banks, insurance companies, railroads and shipping interests dominating the Bay Area. In particular, University of California expansion which drives up taxes should be stopped and small homeowners should no longer pay property taxes. We will demand a direct contribution from business, including Berkeley's biggest business—the University, to the community until a nationwide assault on big business is successful.

10. We will defend ourselves against law and order. America's rulers, faced with the erosion of their authority in Berkeley, begin to take on the grotesque qualities of a dictatorship based on pure police power. We shall abolish the tyrannical police forces not chosen by the people. States of emergency, martial law, conspiracy charges and all legalistic measures used to crush our movement will be resisted by any means necessary—from courtroom to armed struggle. The people of Berkeley must arm themselves and learn the basic skills and tactics of self defense and street fighting. All oppressed people in jail are political prisoners and must be set free. We shall make Berkeley a sanctuary for rebels, outcasts and revolutionary fugitives. We shall attempt to bring the real criminals to trial; where this is impossible we shall implement revolutionary justice.

11. We will create a soulful socialism in Berkeley. The revolution is about our lives. We will fight against the dominating Berkeley life style of affluence, selfishness, and social apathy—and also against the self-indulgent individualism which masquerades as "doing your own thing." We will find ways of taking care of each other as comrades. We will experiment with new ways of living together such as communal families in which problems of income, child care, and housekeeping are mutually shared. Within the Berkeley movement we will seek alternatives to the stifling elitism, egoism, and sectarianism which rightly turns people away and creates organizational weakness. We have had enough of supposed vanguards seeking to manipulate mass movements. We need vanguards of a new type—people who lead by virtue of their moral and political example; who seek to release and organize energy instead of channeling or curbing it; who seek power not for themselves but for the people as a whole. We firmly believe in organization which brings out the leadership and creativeness existing in everyone.

12. We will create a people's government. We will not recognize the authority of the bureaucratic and unrepresentative local government. We will ignore elections involving trivial issues and personalities. We propose a referendum to dissolve the present government, replacing it with one based on the tradition of direct participation of the people. People in motion around their own needs will become a decentralized government of neighborhood councils, workers councils, student unions, and different subcultures. Self-management in schools, factories, and neighborhoods will become commonplace. Locally chosen "people's mediators" will aid those desiring to settle disputes without referring to the illegitimate system of power.

13. We will unite with other movements throughout the world to destroy this . . . racistcapitalistimperialist system. Berkeley cannot be free until America is free. We will make the American revolution with the mass participation of all the oppressed and exploited people. We will actively support the 10-point program of the Black Panther Party in the black colony; all revolutionary organizing attempts among workers, women, students and youth; all Third World liberation movements. We will create an International Liberation School in Berkeley as a training center for revolutionaries.

We call for sisters and brothers to form liberation committees to carry out the Berkeley struggle. These committees should be small democratic working groups of people able to trust each other. We should continually resist the monster system; our emphasis should be on direct action, organizing the community, and forming a network of new groups. Together as a Berkeley Liberation Movement, the liberation committees will build people's power and a new life.

Denis Hayes:
Earth Day: A Beginning

Articles on ecology generally tend to lead off with lists of disasters. But the shock effect of disasters is gone. Today such lists may even be counterproductive. They suggest we have a number of specific problems we must address. We don't. We have The Problem. All ecological concerns are interrelated parts of the problem of perpetuating life on this frail planet, and our approach to them must be holistic. It is absolute folly to continue to pursue piecemeal solutions—when we know full well that the pesticides, the detergents, and the dams are all fouling the same river.

From *The Progressive,* April 1970. Reprinted by permission of the publisher.

This is not to say that the new ecologists *oppose* patchwork improvements—only that we're fairly indifferent to them. If bandages and baling wire make life a little better, that's fine. But the cosmetic alterations being offered by our politicians and our industrialists don't really speak to The Problem at all. They are the kinds of marginal compromises that a skillful player makes to keep control of the game. The precedents are clear.

Other social movements have tramped across the dusty American stage. Many began in search of fundamental change; all failed. Our movement must be different.

Until recently, American movements tended to have a vulnerability. Relying heavily upon an economic analysis, they tended to focus at least in part upon material goals. And this made them vulnerable. The American economy can manufacture wondrous quantities of goods. So if a militant group wanted a piece of the pie, and was willing to fight for it, the economy would simply produce a little more pie and give it some. And with its material goals addressed, the movement invariably lost its teeth.

The contemporary revolution, however, does not rely upon an exclusively economic analysis, and its goals are not acquisitive. It will be impossible to "buy off" the peace people, to "buy off" the hippies, to "buy off" the young Black militants, to "buy off" the ecology freaks.

We can't be bought, because we demand something the existing order can't produce. We demand a lower productivity and a wider distribution. We demand things which last, which can be used and reused. We demand less arbitrary authority, and more decentralization of power. We demand a fundamental respect for nature, including man—even though this may sometimes result in "inefficiency."

Dissident groups accentuate different concerns, but our fundamental goals tend to be shared: ending exploitation, imperialism, and the war-based economy; guaranteeing justice, dignity, education, and health to all men. A focus on one concern does not mean a neglect of the others: We are able to seek more than one goal at a time. Those of us who have fought against the war will continue to do so until it is ended; those who have sought racial justice will not be satisfied until it is realized.

All these goals fall under a single unified value structure. This value is difficult to articulate, but posited most simply it might read: "the affirmation of life." This is a clear contradiction of most things for which America stands.

America is the New Rome, and is making Vietnam the new Carthage—razing her villages to the ground and salting what remains of her fields with long-lasting defoliants. America is the new Robber Baron—stealing from the poor countries of the world to satiate a gluttonous need for consumption. We waste our riches in planned obsolescence, and invest the overwhelming bulk of our national budget in ABMs and MIRVs and other means of death. Each of our biggest bombs today is the explosive equivalent of a cube of TNT as tall as the Empire State Building. And the only reason we don't have bigger bombs is that we don't have a means of delivering bigger bombs. Yet.

At the same time we are systematically destroying our land, our streams, and our seas. We foul our air, deaden our senses, and pollute our bodies. And it's getting worse. Our population and our rate of "progress" are both expanding geometrically. Tens of thousands could die in Los Angeles in a thermal inversion which is now probably unavoidable—and not one element of the existing system of air pollution control can do a thing to reduce the flow of poisonous traffic into that city. That's what America has become, and that's what we are challenging.

America has become indifferent to life, reducing that vibrant miracle to a dead statistic. This callousness has allowed us to overlook the modest, intermediate consequences of our crimes. If 50,000 people are killed, if ten million people starve, if an entire country is laid waste—we have learned to tuck the information into the proper file and write the affair off as a mistake.

We have to "unlearn" that. These aren't mistakes at all; they are the natural offshoots of the "growth generation"—of the neo-Keynesian mentality that *still* expects to find salvation in the continued growth of population and production. Nurtured in our frontier heritage as the short-sighted inhabitants of a bountiful, underpopulated country, this mental set (found in every economic text in our schools) has yet to grapple with the elementary fact that infinite expansion is impossible on a finite planet.

The implications of these old myths in the current setting are enormous. We now have the potential to destroy life on the planet, and in our rapacious plundering we are flirting with some frightening probabilities that we will do just that.

In a society in which death has lost its horror, with the slaughter of three wars under our belts and our streets full of mugging and indifference, a group of people—mostly young—is beginning to stand up and say, "No." We are beginning to say, simply, "We affirm life—a life in harmony with Nature." And that's what April 22 is really all about. . . .

Sometimes, around dawn, we finish the previous day's work and I go for a walk. The sky slowly lightens, and in time the streets become snarled. People jostle around me, rushing from their dissatisfied homes to their unhappy places of labor. And they seem to know something is wrong. The Blacks, and the Wallace working class, the great silent majority—all feel the tense unease.

What's wrong is that the species is breaking the most fundamental biological rule; much of it seems unwilling to fight for its life. At a time when survival itself depends upon the development of an ecologically balanced world, the worshippers of "progress" are instead embracing death.

The inconsistencies are becoming clearer, and a whole generation is rapidly losing its naivete. A pregnant new politics is developing, which cuts across former boundaries and which scorns the sterile inflexibility of existing institutions.

We are mapping a struggle—not only against the vested interests of the giant corporations, not only against the paid-off senility in our Congress, not only against the Strangeloves in the Pentagon. Survival demands something more. The very survival of the species has come to demand an ecologically-balanced planet—a state at variance with most of the value assumptions of Western civilization. Our challenge must consider not only the pieces but the whole.

The course ahead is dangerous and complex and may be impossible. But new bonds of brotherhood are rapidly linking diverse people around a profoundly simple value: the affirmation of life. And with intelligence, courage, and luck, we may win.